AF324336

Worldviews, Science and Us
Interdisciplinary Perspectives on
Worlds, Cultures and Society

editors

Diederik Aerts
Vrije Universiteit Brussel, Belgium

Bart D'Hooghe
Vrije Universiteit Brussel, Belgium

Rik Pinxten
Ghent University, Belgium

Immanuel Wallerstein
Yale University, USA

Leo Apostel Center, Brussels Free University,
August 2005; July 2007; September 2010

Worldviews, Science and Us

Interdisciplinary Perspectives on
Worlds, Cultures and Society

World Scientific

NEW JERSEY · LONDON · SINGAPORE · BEIJING · SHANGHAI · HONG KONG · TAIPEI · CHENNAI

Published by

World Scientific Publishing Co. Pte. Ltd.

5 Toh Tuck Link, Singapore 596224

USA office: 27 Warren Street, Suite 401-402, Hackensack, NJ 07601

UK office: 57 Shelton Street, Covent Garden, London WC2H 9HE

British Library Cataloguing-in-Publication Data
A catalogue record for this book is available from the British Library.

WORLDVIEWS, SCIENCE AND US
Interdisciplinary Perspectives on Worlds, Cultures and Society

ISBN-13 978-981-4355-05-6
ISBN-10 981-4355-05-4

Printed in Singapore by World Scientific Printers.

CONTENTS

INTERDISCIPLINARY PERSPECTIVES ON WORLDS, CULTURES AND SOCIETY

DIEDERIK AERTS AND BART D'HOOGHE

Center Leo Apostel for Interdisciplinary Studies,
Vrije Universiteit Brussel, Belgium
E-mails: diraerts@vub.ac.be; bdhooghe@vub.ac.be

RIK PINXTEN

Center for Intercultural Communication and Interaction,
University of Ghent, Belgium
E-mail: hendrik.pinxten@UGent.be

IMMANUEL WALLERSTEIN

Department of Sociology,
Yale University, USA
E-mail: immanuel.wallerstein@yale.edu

This volume is part of the 'Worldviews, Science and Us' series of proceedings and contains several contributions on the subject of interdisciplinary perspectives on worlds, cultures and societies. It represents the proceedings of several workshops and discussion panels organized by the Leo Apostel Center for Interdisciplinary studies within the framework of the 'Research on the Construction of Integrating Worldviews' research community set up by the Flanders Fund for Scientific Research. Further information about this research community and a full list of the associated international research centers can be found at `http://www.vub.ac.be/CLEA/res/worldviews/`

The first contribution to this volume, by Koen Stroeken, is entitled 'Why consciousness has no plural'. Stroeken reflects about the way philosophical questions of epistemology influence theories of anthropology. More specifically, he focuses on how the notion of 'spirit' plays an important role in many cultures studied by anthropologists, analyzing how different elements of worldviews, but also specific aspects of modern physics, can lead to original hypotheses on this notion.

The next contribution, by Hendrik Pinxten, is entitled 'The relevance of a non-colonial view on science and knowledge to an open perspective on the world'. Pinxten argues that the so-called 'Methodenstreit' was wrongheaded, because both camps were guilty of the colonial attitude. They failed to take into account the perspectives on knowledge and other traditions in any genuine way, let alone allow for conceptual openness. Pinxten scrutinizes the conceptual and epistemological problems involved as well as the methodological issues of praxiology and performance theory.

The third contribution is by Jeroen Van Bouwel and is entitled 'An atlas for the social world: What should it (not) look like? Interdisciplinarity and pluralism in the social sciences.' Van Bouwel's aim is to examine how an interdisciplinary approach may produce an atlas to help navigate the social world. For this purpose, he looks into different social theories to evaluate how they might work together or start a mutual dialogue, after comparing the features of theories and maps, and relying on the different strategies that have been defended for interdisciplinarity in the social sciences. He illustrates his approach with actual proposals made by, *inter alia*, World-Systems Analysis, Critical Realism and Economics Imperialism. He then makes a case for question-driven interdisciplinarity, illustrating his argument by discussing recent developments in economics, i.e. the debate between the orthodox and heterodox theories, the pleas for pluralism, and the impact of globalisation — and related institutional developments — on economics as a discipline.

Patricia Frericks and Robert Maier authored the next contribution, entitled 'Worlds of legitimate welfare arrangements: A realistic utopia on pensions'. Frericks and Maier analyze several pension-determining factors by studying different European pension reforms, presenting an original evaluation of pension policies in the European Union. Their approach is 'realistically utopian', in the sense that it combines recent political, social and economic reform considerations with normative and theoretical ideas. Their aim is to answer the question of whether there are ideal ways of combining these factors with the ultimate goal to outline a legitimate and sustainable pension system.

Hans Alma and Adri Smaling's 'Imagination and empathy as conditions for interpersonal understanding in the context of a facilitating worldview' studies the notion of empathy and its relationship with worldviews. The authors conceive of empathic understanding as a two-dimensional concept. The mental dimension refers to affective, cognitive and interpretive facets or phases of empathic understanding and the social dimension refers to expressive, responsive and interactive facets or phases of empathic understanding.

These two dimensions are intertwined and the optimal form of empathic understanding is called 'dialogical-hermeneutical empathic understanding'. Furthermore, the importance of imagination and its development for optimal empathic understanding is elaborated.

The next contribution is from Lieve Orye, entitled 'Worldview as relational notion? Reconsidering the relations between worldviews, science and us from a radical symmetrical anthropology'. Lieve Orye develops an anthropological reflection on the notion of worldview. Highlighting the commonly established connection between worldview and map, orientation and globality, as well as the frequent discussions in Christian and other philosophical circles on the worldview-related notion of relativism and the swaying movement between conscious commitment and unconscious bias, she discusses the possibility of the notion used in such reflections being problematic. On the basis of Ed Hutchins' work on ship navigation and the reflections of two reviewers of the latter's work, viz. Bruno Latour and Tim Ingold, the author points out a foundational error in human sciences and introduces the possibility of considering the notion of worldview as a relational notion.

Richard E. Lee's contribution is entitled 'The structures of knowledge in a world in transition'. Richard Lee considers how the structures of knowledge, the separation of facts from values into the two cultures of what eventually would come to be called the sciences and the humanities, emerged as fundamental components of the modern world-system along with the axial division of labor and the interstate system in Europe in the long sixteenth century. He argues that the restructuring that started in the late nineteenth century and resulted in the creation of the social sciences between the poles of the sciences and the humanities, is now in crisis. He investigates how developments in complexity studies and cultural studies put into question the utility of not only the presently accepted disciplinary boundaries, but also the attendant received epistemological, methodological, and theoretical approaches.

This is followed by Ellen Van Keer's 'On bridging theory and practice in the perspective of history'. Ellen Van Keer reflects about possible bridges between theory and practice in historical studies. She argues that, while traditionally approaches have been evidence-based, focused on practical and empirical matters and largely ignored problems of interpretation and theory, 'postmodernism' has brought to light that non-empirical factors are inevitable and pose major epistemological obstacles in the perspective of history. In particular, it has become clear that historians cannot but translate the past into their own words and concepts, and should be wary of hidden linguistic and cultural factors that actively help shape the knowledge they produce. To deal with this problem, Ellen Van Keer proposes a critical examination of the roles and concepts of (1) epistemology, (2) theory, (3) philosophy, (4) methodology, and (5) historiography in the study of history. Historiography emerges as a most promising subject and field to expose and address issues of theory in the practice of writing history. Furthermore, (6) archaeology, while traditionally describing the empirical-analytical study of past material objects, has recently also been developed into a critical-theoretical concept that exposes hidden conceptual foundations, implicit modern assumptions, and intrinsic linguistic structures determining our knowledge about the past. Thus, it integrates the main opposing and complementary tenets and approaches in the 'new' perspective of history.

The final contribution is by Gert Goeminne, Filip Kolen and Erik Paredis and is entitled 'Addressing the sustainability challenge beyond the fact-value dichotomy: A call for engaged knowledge'. Goeminne, Kolen and Paredis suggest we should stop conceiving of scientists and politicians as mutual antagonists using their own specific weaponry, and instead begin to regard the place of their encounter, not as an arena, a battle stage, but as an agora, a meeting place where experts and lay people alike can join in the shared practice of 'engaged knowledge'. The authors argue that, to promote the agora as a political space for addressing sustainability issues beyond the fact–value dichotomy, three main tasks are to be addressed. First of all, at the conceptual level, space will need to be opened up to host the agora. If we start thinking in terms of the co-constitution of subject and object and a conception of science as an activity that is necessarily engaged, this will allow us to explore the area in between the subject–object poles, rather than taking biased stances in favor of either side, be this science/policy, fact/value or knowledge/power. Secondly, it should be made clear what politics will look like in the agora. Participation will acquire a different status, characterized by its 'engagement', with knowledge practitioners, lay and expert, joining in the same practice. Their acts and statements can be checked against a set of measures constituted from within the dynamics of this common practice, in response to what they recognize as what is at issue at any time. Thirdly, the agora needs to be put at work. The authors therefore look into concrete policy approaches in the broad field of sustainable socio-technical transitions as potential entry points for their agora-oriented approach to engaged knowledge.

WHY CONSCIOUSNESS HAS NO PLURAL

KOEN STROEKEN

Institute of Anthropological Research on Africa
University of Leuven (K.U.Leuven), Belgium
E-mail: koen.stroeken@soc.kuleuven.be

Anthropologists respecting the cultures they study should seriously engage with their epistemologies, for instance regarding the principle of extra-natural agency called spirit. Without having to embrace the historically determined beliefs, we may recognize behind cultural differences certain underlying structures of meaning, or worldviews, which cross-culturally recur and subtend theories in disciplines as wide apart as sociology and physics. The paper reports, almost in a stream of consciousness, how an anthropologist initiated among spirit mediums might perform an exercise in 'anthropic quantology'. Spirits stand for a notion of contingency more subtle than our 'chance versus determination' from which the paradoxes of quantum mechanics and consciousness seem to stem. We explore a radical inter-disciplinary option called physical dualism. It equates meaning structures with the laws of nature, and defines the latter as a second type of matter, invisible yet energizing the mind just as spirits. Consciousness would be the 'cultural selection' to which individual bodies have unique access. While quantum mechanics seems to deal with the more encompassing reality of naturally selected 'many worlds', Einstein's spacetime may be expressing this one anthropic selection called consciousness.

1. Introduction

The backdrop to this chapter is an anthropologist's discomfort with the dichotomy between scientific knowledge and the kind of beliefs he or she is dealing with in the field. As Gellner argued,[1] Western thought cannot help opposing reason and culture, knowledge and belief. Other societies allow for a continuum between the two, so the anthropologist may wonder whether the dichotomy is not characteristic of our society institutionalizing science and religion. Why indeed is it so hard for modern thought to go beyond dichotomies such as Darwinism versus Intelligent Design? Few scientists will deny that religion follows from natural selection. Yet, they will have trouble admitting that the same holds true for science. It seems logical that natural selection played a role in shaping the human brain, conditioning scientists to perceive only a certain part of the world. If we take evolution seriously, should the unperceived part then not be anticipated in non-scientific, even religious explanations to further develop brain and mind? Is this not, after all, how the boundary with religion has continued to shift? Strangely, that reciprocal position, perfectly consistent with evolution theory, is discouraged of late. Instead, segregation rules academics. The most basic dichotomy is, since Dilthey, that of nature and culture. Segregation becomes colonization as influential thinkers such as sociobiologist Dawkins and physicalist Dennett apply Darwin's theory to reduce everything human to perceptible biological processes. Are thoughts equal to neural circuits because only the latter can be scanned?

Conspicuously absent in physicalism (also known as materialism) is, paradoxically, the most prominent physical theory. Quantum mechanics shows how to know things that cannot be directly perceived. Quantum theories converge with the anthropological caution that perception (via sensory equipment not unlike that of other mammals) does not seamlessly emanate into explanation (meanings). Culture intervenes, even in the language used by natural scientists. More so, the fact that people despite their cultural and biological limitations can transcend perception and actually get to know the world, in whatever way, is precisely what has led many cultures to assume a 'connection' (indeed *religio*) with something else than the natural environment. This paper explores the possibility of such a second environment, in light of unresolved questions of both mind and matter. The alternative to nature's chance is not divine will. To transcend this dichotomous thinking, we must reinterpret contingency in light of the selection of laws that made nature possible and continue to guide its self-organization. When scientists, artists and their ancestors have attempted to capture the laws of the world are they not really after basic meanings? Meaning is the kernel of mind, but comes not from our perceptual apparatus. Could nature's selection of laws be the direct source of meaning? The paradox is that our species developed this very selection over millions of years. We delimited the universe we can be conscious of.

Can we grasp a notion of contingency more subtle than our 'chance versus divine will'? I think to have encountered it among the Sukuma farmers and herders with whom I lived for two years. They live southeast of Lake Victoria, in the so-called cradle of mankind. Not that this matters. Westerners have long disparaged, on the grounds of materialism and Darwinian evolution, any cultures hinting at ancestral spirits, divinities, or the old intermediate theory, that of Plato's eternal ideas. I say 'hinting' because institutionalized, sanctioned beliefs, as in Christianity or Islam, are uncommon in the societies anthropologists mostly study. If we disconnect religion in the broad sense from its usage by institutions such as the Church, might we not be discerning in these everyday beliefs a wise, cautious stance about meaning not necessarily following from perception? Let's take an example closer to home. Cultural norms generally revolve around social equality and protection of the weak. They seek to curb instinct and turn natural selection on its head by adapting the environment to fit ideas whose origins are largely unknown. Is the suggestion then not again that humans, by being conscious, draw on a source that parallels biological and environmental impulses? And is our aversion against such invisible, second causality not why we as scientists will retort that those norms serve natural selection anyway; and why we might downplay the human capacity of actually knowing processes such as these and thus of altering their course? At the same time, though, we lament the destructiveness of the human species to others and to itself. We thus fear that our knowledge, harmless in itself, becomes a danger in combination with our animal drives. But precisely because we secretly intuit that the source of this knowledge does not coincide with our biological constitution, we do not lose hope and expect a solution to lie in store.

Cultural differences cannot conceal that a limited number of structures reappear in meaning-making. Sophisticated specializations in science cannot conceal the recurring structures either (e.g., 'compensation' in thermodynamics, economy and witchcraft beliefs). They are more basic than mathematics. What is the status of those meaning structures? The assumption has been that they simply emerge from natural micro-events, without having an actual influence. Culture would emanate from nature; be implied in the big bang as it were. I propose instead that nature pulsates with its condition of possibility, namely the selection from which it came into being, and that this is the level of culture, something preceding nature. Quantum experiments reveal that possible positions of particles exist next to each other, interpretable as parallel universes that can be indirectly perceived. I argue that human consciousness consistently selects one universe, pre-determined and thus non-local. That selection constitutes the source of our consciousness. That selection is the very condition of possibility for our universe and it explains why humans can know that condition at all in scientific laws or artistic and religious articulations. Rather than Cartesian 'thinking stuff' (*res cogitans*), I propose a second type of matter which co-originates with the first type and can interact with it (like a wave affecting neural circuits). I call it culture, in the singular. It prefigures the plural 'states' of culture studied by ethnographers.

Culture has a causality of its own, interacting with nature's causality. My strongest argument for this physical duality is the materialist contradiction which persists as long as we do not accept the duality and which suddenly vanishes once we do. In physics, the contradiction goes as follows: every event in the universe is indeterminate (uncertainty principle, quantum wave function), yet for the universe to exist with its properties (non-local spacetime, quantum entanglement) every decision must have been made in advance. Significantly, this paradox of matter has its equivalent in the human sciences as the paradox of mind: human consciousness is characterized by freedom (volition) and yet for it to exist with its properties (a system to which all meanings relate) every meaning must be included in the system. Or: thoughts are mine, yet they arrive from somewhere I cannot control. The paradox strongly reminds, we will see, of Chwezi mediums who attribute the content of their consciousness to spirit presence. If duality is preferred over contradiction, the Chwezi ritual of initiation offers a solution. It turns out that something can be fixed yet free if it exists in two realms. Without the necessary shift of perspective, there will be conflation and thus contradiction. I distinguish the two realms as follows. Considered in themselves, my thoughts are free. When considering the body and environment they are part of, they are not. Considered in itself, the universe is free (natural selection). But once we accept that there is a realm beyond our universe where it exists as one next to other universes, then we realize that the selection marking our universe is fixed in relation to those others, thus at a level beyond space and time (cultural selection). In analogy, the chair we are sitting on is static, while at a lower level it is dynamically formed by atoms, each of these again appearing fixed despite their internal chaos. The choice seems between limiting ourselves to the visible world, and winding up in contradiction, or accepting the basic duality proper to an invisible layer.

2. Towards an Anthropic Quantology

Anthropology is unique as it has become a discipline concerned with proving that it is not doing what it is entitled to: studying 'the human', *anthropos*. The discipline has gradually split itself up in two branches, social and biological, which cover supposedly separate dimensions of being human. The social branch dismisses any claim about the human because of its universalism, its denial of cultural specificity. Hence, when talking of culture, social anthropologists today emphasize the plural: cultures. Modernity becomes modernities.[2,3] This conceptual preference is understandable if we consider the atrocities in the last century done under the label of the human. Deviants from the imagined ideal were colonized, segregated, gassed. Social scientists understood that the human is always a construct and that once the label is reified, its universalism can be dangerous. The benign but still devastating version of universalism — ethnographers in the field came to realize — is the spread of Western doctrines such as Christianity, Enlightenment and perhaps the contemporary idea of 'development'. These pretend to usher the ignorant individual into the realm of truth, at whatever cost in terms of intricate balances of collective well-being that local communities may have developed overtime. Positive sciences operate on the assumption that their success and technology — or in terms of natural selection: the success of their fit to the environment — attests to their global truth, no matter how impoverished and socially, psychologically and ecologically disturbing this worldview appears to be. In brief, anthropologists have every reason to insist, not just on cultural diversity, but on hard cultural difference, in all its fascinating and perplexing incommensurability.[4] Culture has to be read in the plural.

Yet, in the wake of this postmodern paradigm, whose relativist inclination imbued the sciences at large and hammered 'cultural difference' into public discourse, something unintended happened. While anthropologists cut the human apart and prefer to call themselves Africanists or so, the constant reference to cultural difference is increasingly being abused in Europe to justify and revive rightist ideas on the impossibility of intercultural affinity. The repressed (once again) returns with a vengeance. It turns out that the anthropological label of the human has never been the problem. On the contrary, the mistake has been to conceal what unites humans. Each of us different, yet not entirely — what disciplines can lead the way?

Biology has been helpful. Genetics clearly refute the idea of race to describe human variation.[5] Nevertheless, while each individual looks different, family resemblances can always be found on regional to planetary scale to isolate categories. But how relevant is genetic identity for humans? Sharing 99% of our genes with chimps may prove to the biologist how close we are to apes, but to others it may prove that biology does not tell the whole story about humans. If not biology, then what would? Could 'culture' realize its vocation as complement to 'nature'? Every fieldworker will remember the surprise, long after the so-called culture shock, that as a European visitor living for two years in a village where the West lies in the margin and where a new language and conduct has to be learned, one can nevertheless grow very close with one's hosts and share experiences no less intensely than with one's best friends back home. Amidst difference emerges similarity. What status does this similarity have? Thinking of everyday situations with my host Paulo Makufuli, shared expectations and simultaneous sighs of relief pointed not to a mimicking of bodily technique but to similarity in feelings. However, as far as empirical science can tell, the content of our experiences could never have been the same. Let's face it: the experiences we accumulated in a personal past were entirely different. The bodies by which we perceive the environment never came close to coinciding. Not to mention the neural circuits which people's thoughts are supposed to be made of: not one neuron is shared. What then is?

A mechanistic answer is that the neurons in you and me are not the same but alike; they follow the same laws and thus produce similar psychological reactions or states of consciousness in a given environmental context. The similarity between our feelings — the totalities produced from those parts — may very well be mere appearance, a formal likeness only. Therefore we can never really know another person's experience (what you call green may be very different from the green I always perceive) — let alone know the experience of people from another culture. A corollary of the mechanistic worldview is cultural relativism. Far from combating the materialist idiom (as often believed), mainstream relativism grants materialism's reign in things not cultural and thus consolidates Dilthey's old scientific segregation of nature (causality, explanation) and culture (intentionality, interpretation).[6] Violations of the segregation are risky; not to be taken lightly. Dualists combat materialists in the mind/matter debate, but without rejecting the segregation, or even the mechanistic model. As proposed in the 'property dualism' of Chalmers and others,[7] the content of consciousness would consist of subjective qualities coined 'qualia'. These non-physical entities would parallel our brain's activities and build up the content of mind. Such non-physical realm perhaps annoyed materialists but never disconcerted the natural sciences. That would only be the case if consciousness would involve a non-biological causality, not directly perceptible but manifest, thus topping the brain and its environmental impulses. How to imagine this other type of causality which would establish genuine interdisciplinary research (but which the phenomenologist's concept of intentionality excludes)? Anthropologists have for over a century been discerning structures across cultures. I will argue that these structures refer to the cultural selection marking the human species. However, the dominant view is, in keeping with the academic segregation, that the structures only have a reality in the macro-perspective of 'the social system'. (A variant, equally far from disconcerting, is Bloch and Sperber's naturalist position that cultural patterns are emergent properties of micro-events interacting.[8]) Durkheim likewise spoke of social facts. The problem, haunting the social sciences until today, is that these facts depend on the author's pretension of a knowledge transcending the social. Eventually, scepticism appeared the logical position from Habermas to Rorty: our knowledge lacks final ground and is grounded by power structures or, at best, social consensus.[6]

There is something ambiguous about this position, since it never stopped anyone from perfecting social theories and being convinced of the corrections. Furthermore, were we to disregard Durkheim's segregation, we would realize that our brain is shaped by natural selection. Not only do our thoughts, as Kant speculated, differ from what is really happening out there. We detect those sensory impulses that have since long benefited the survival of our species. Does our brain then not select from the environment what is good to think rather than what is true? The cross-cultural similarity between thought patterns seems to confirm the natural selection. Still, when it comes down to it, scientists, artists, cult novices or cave painters feel they unravel something true. We know because we perceive, the scientist contends. But concepts do not emanate from perception. A claim about the world is something imperceptible. This possibility which we call meaning-making, is all the more remarkable if despite cultural and biological determinations our symbolizing would uncover something universally valid. The standard reply is that accurate percepts of the world gradually developed into accurate concepts of the world. But then we should come across a transitional language, or an 'almost' conceptualizing species. Anthropologists have looked for these among so-called primitive tribes and found none.[9] A complete system seems to exist from the start, with varying degrees of complexity developed over time. If the biological evolution towards meaning-making was gradual, the development of culture seemed more like a 'big bang' once the access to meaning was gained. The reason, I contend, is that the meaning system has an influence in its own right. Just as our perceptual apparatus slowly evolved while the natural environment itself fires all its impulses at once, culture refers to a second type of environment, not one to perceive, but 'to know'. It tallies with the anthropologist's experience that the culture he or she grew up in is permeated with the same old ancestral 'spirits' which in other cultures also energize the mind. Our bodies and neural networks make a difference. But what they differentiate

12

is one system of meaning, one source of consciousness proper to our species.

The advantage of this access is double. Hanging onto concepts, feeling meanings, allows us to suspend homeostatic response and stand or cultivate emotions. That was the role of natural selection. The unintended side-effect of this access has been that we get thoughts whether or not they are good for us (as we very well know when stumbling upon a truth destructive for ourselves or others). The only way we could leap from perceptual encapsulation to conceptual understanding, and go on in the way thoughts do, was to get attuned to an imperceptible yet physical world that is closely related to the natural environment we live in. Drawing on quantum theories dealing with matter that cannot be directly perceived, I thus propose a physical dualism, which side-steps the dichotomy of physicalism versus dualism that rules the mind/matter debate. I submit that the human mind parasites on the laws that physically exist to delimit our universe. As Lévi-Strauss inverted Kant's terms, the mind is structured like the world.[10] That is why the laws of the universe are not found by hyper-wiring ourselves but on the contrary by weeding neural excess so as to have our thoughts dominated by those very structures underlying the world. Moreover, without such culture in the singular, determining what all cultures can variedly engage in, there could be no such thing as anthropology, one idiom talking meaningfully about cultures. There are modes of culture, just as there are modes of consciousness. To talk of 'consciousnesses' not only sounds strange. It can only refer to intelligent life in other universes than the spacetime selected by our species. Human consciousness has no plural.

But here's the rub. The spacetime our species is determined by, appears to be the one it has selected! (In cosmological terms, Einstein's spacetime universe is a closed sphere, but expanding, changing). Cultural selection is not fixed by some divine creature, but by ourselves, by all cultures, of all times. How to figure out this idea, reminiscent of a snake biting its back? I will call in the help of the age-old Chwezi cult, which developed in interlacustrine Africa, independently from massive civilizing traditions such as Christianity, Hinduism and Islam.

My study lies at the cross-roads of three concerns: balancing the current hyper-focus on difference, on nature and on culture. The following principle weds the contemporary interest in cultures to the old anthropological quest for human nature: to find the fundamental difference between two ways of thinking is to find a level of analysis at which to speak of both. Such 'mediated theory' looks for the variable encompassing two values. A mediated theory does not envisage an alternative or the kind of anti-scientific paradigm-shift which romantic counter-movements have been propagating in the slipstream of Enlightenment. It approaches their holistic worldview like an anthropologist internalizing a different culture in order to verify what the mechanistic model he or she has been trained in, lacks to possess that explanatory value too. Accordingly, this paper on consciousness will proceed in two steps. First, I give a brief ethnographic illustration of the Chwezi holistic worldview. I discuss its relation to quantum theoretical understandings of matter and consciousness. The key-concept is selection (from a given whole). Second, I discern a mechanistic model in the Chwezi rituals, which can serve to recast the holistic framework in positivist terms. The key-concept is influence. We discover how spectacular the holistic view is in mechanistic terms.

3. Meaning is Matter

My concept of consciousness was transformed by extensive participation in 1997, 2000 and 2003 in the Chwezi cult which since long before colonization has been initiating novices from Interlacustrine Africa into spirit possession and mediumship.[11] Spirit mediumship means access to a physical being of another order. The sacrifice or suspension of self is akin to shamanism, deep meditation, and mystical experience. Far from proving that perception of the spirit actually takes place, what interests us here is the structure many cultures have opted for to make perception of something external happen. The mystery of how humans, as opposed to registering machines, can be conscious of something is comparable to the enigma of the Chwezi being filled with external presence. We know that human perception works but not how matter does it. Conversely, about the Chwezi we can never be sure whether the spirit is really perceived, but we know how they do it according to the ritual. This section tries to complement both kinds of knowledge.

The method which the Chwezi ritual employs for bodies to give passage to the spirit, operates on the assumption that there is something out there giving away energy. The challenge is, quite mechanically, to capture the energy. To do so, however, no magic or technology need be activated. Rather, the dominant impulses of mind have to be removed as they obstruct that other influence. This is why the whole ritual of initiation consists in defying the dominant distinctions of society, down to the universal prohibition of incest. For instance, the rules on gender and age no longer apply. Novices have a parent designated who is to (symbolically) copulate with them. Symbolic play converts deeply socialized feelings into abstract signs of inclusion and exclusion. Novices become aware of a discontinuity between what they feel and the meanings that have been taught to them. Socially induced feelings of synchrony (good) and intrusion (bad) in relation to external impulses become meanings (inclusion/exclusion) so that feelings can be explored of another order. The binary code of inclusion/exclusion is how sociologists define the meaning system.[12] In the Chwezi rite we thus witness an age-old procedure magnified: meanings, which constitute culture, serve to make abstraction of feelings (experiences) in order to explore yet other feelings, which can very well be feelings concerning the abstracted feelings that became meanings (culture is a stimulant itself rather than a mere translator of natural events). In other words, consciousness exists in the (quaternary) flow of us feeling (binary) meanings.

The contrast is intended with Damasio's neurobiological definition of consciousness as 'feelings of what happens', which leaves out the dimension of meaning, thus does not explain how consciousness of the environment is possible.[13] How can external events, besides being registered in order for our bodies to respond, actually obtain meaning and become thoughts? The naturalist position is clear: "[e]volution gave rise to organisms with subjective feelings. These convey significant survival advantages, because consciousness goes hand-in-hand with the ability to plan, to reflect upon many possible courses for action, and to choose one".[14] Or: "Evolution has wired the experience of pain deep into the program to make you pay attention at a very primitive level".[15] Damasio accordingly approaches perceptions as somatically marked associations between stimulus and response.[13] But here appears the limitation of the neurobiological perspective: somatic marking refers to the homeostatic function alone, to the search for survival. Perception requires the body to momentarily do the opposite and accept the disturbance, which is a danger to survival. This principle can be recognized in basic life-forms. A snail has sacrificed the bliss of ignorance for a perception of poking, which raises its survival chances. Every little poke of my finger fires a bunch of neurons in the snail's brain, because its organism has acquired some form of synchrony with the intrusion. Quite tellingly, in this the human organism is reputed to have gone far, as reflected in our parasympathetic nervous system suspending fear-responses.[16] Whether it was drastic climatic change or interspecies war that drove the closely related Neanderthal to extinction, one thing is certain: our Pleistocene ancestors overcame these extreme ordeals.[17] Dramatically raised levels of anxiety led our species to perfect neural strategies to retain feelings of unbearable intensity. The survival advantage of meanings is that they are located outside the body and thus free from somatic marking. A concept such as danger allows the body to register the emotion or consider the object of emotion, and keep it close to mind, without having to directly react by fleeing or repressing it. Typically, we preclude human accountability for moments when such access to the meaning system is missing, as in blind fear.

Where do the meanings reside? They have, as yet, not been detected in the brain.[18] In fact, if we look at how abstraction strategies continued to evolve, it appears they all work through pulsation with something external: when thoughts did not sufficiently detach us, they were spoken as words; when words still engaged us too much, they became written signs. Moreover, the Lascaux frescoes 30.000 years ago, the fractal patterns on pots and clothes, the invention of writing: all translate four-dimensional experiences into two-dimensional representations of meaning. Does this not indicate that the meanings, and their system, are of another order? In fact, would these meanings in art and science make any sense if we did not believe that rather than figments of the brain they are actual entities energizing our consciousness? In the Chwezi procedure we also find that a higher consciousness, or synchrony with the spirit's intrusion, is based on the suspension of somatic response (Believe it or not, those spirits are also represented as dimensionally halved: they have one eye, one arm, one leg). The question is what flows in at suspension. Could the human body have built the passage to another universe than the one it belongs to? Is there any other way to bridge the gap between matter and mind than accepting the latter as another type of matter pouring in?

Chalmers called the gap the 'hard problem' of consciousness: how could our hued sensations arise from this grey blob of brain?[7] Smart hybrid terms have been invented to conceal the explanatory gap as in Damasio's 'neural maps of emotional states'. Another example is Lewis-Williams who attributes the similarities of cave paintings by prehistoric people and by San bushmen to 'entoptic designs' structured in the brain and popping up visually under stressing conditions.[19] Obviously, the inner perception of such a design (for instance, the spiral, an image often encountered in cave paintings) says nothing about the meaning it is supposed to have (*in casu*, drowning). A dog too may be visualizing the spiral, but before associating it with drowning, a lot must happen. The associative brain naturally produces percepts, but not abstracts or concepts. There is no smooth transition to inventing or receiving the latter. Natural selection does not fill these in. Nor does the natural environment. We imagine the content of perception to be energized by light, sound waves and so on. But where does the content of concepts come from? How could they be 'derived' from perceptions? Crick and Koch's interest in meaning yielded disappointing Neural Correlates of Consciousness which just as Damasio's somatic markers reduce meaning to the brain's evolutionary in-built associative capacity.[14,20] From those associations to concepts is a big leap. It is one thing to stretch and cultivate the time-span between stimulus and response, but still another to be able to fill the picture up with meanings, referring to positions in a symbolic order.

The Chwezi ritual basically taught mediums how to lose and regain consciousness. Its principle, it seems to me, is that of consciousness itself: meaning-making animals synchronize with intrusion through inclusion/exclusion. The basis of everything is synchrony. Particles colliding can spontaneously generate order.[21] Living beings achieve a dimensional advance by perceiving. The snail synchronizes with intrusion. Natural selection determines which species genetically stored good solutions. The human mind can go further than trial and error as it evolved towards cultural selection. Via inclusion/exclusion, natural laws are experienced at the level of the individual. The laws move from the real world into a thought world. A direct link can be traced from the snail's urge for survival to the individual mind, which may explain why we generally intuit that machines can never have the consciousness which living beings have grown into.

The mystery is how those natural laws made up in our mind could be true, if not themselves energizing the content of our mind. The gradual difference between humans and other animals in terms of cortical sophistication is complicated by this extra-natural environment which our brain is logged into and which decentres human consciousness from the body and its natural environment. We call this second environment culture or the meaning system. It is described in our science, myths, language. For many, culture directly descends from the ancestral spirits (it inspired Plato's eternal truths). Westerners have instead learned to think of it as artificial, corrupting our perception and reasoning.[1] It can be argued, though, that this culturally selected world is the only one which humans can be conscious of, whereas the 'real' world, which our sciences are after, can be no more than indirectly registered. Quantum experiments suggest that the determinate contours we perceive of things in the natural environment should be replaced by probabilities and wave functions. The consequence may be to accept that the second environment, beyond which our species cannot think, is real too. The matter of meaning may exist as a wave synchronizing neurons.[22] Hameroff and Penrose have surmised a Platonic realm to connect mind and matter.[23] It inspired Lutz et al. to scan the neural assemblies of Buddhist monks meditating the concept of compassion.[24] They registered an increased coherence of gamma synchrony (40Hz). Could such physical wave be the matter of a meaning, of an experiential structure generating consciousness?

Mother Nature is physical, causing things to happen. Why not Father Culture? After carrying him to a non-physical realm, the Western fascination grew with cultures where the physical influence from the meaning system is taken to be evident, in the form of the spirit world populated by ancestral truths. Nostalgia, or is there more to it? We say that ideas are 'in the air', and we're in a hurry to share them. The interest is universal in artists, scientists, shamans, in any woman or man telling us what makes the world tick. Perhaps we've known it all along. A meaning system is beaming out constants for thought. On our side there's varied neural reception. To repeat my query about Paulo and I sharing something real in our experiences, is this not the non-neural part in our thoughts that could be shared — humans tapping from one consciousness?

Myriads of bio-psychological and socio-ecological interactions have together generated the writing of this sentence. The chance of another person reproducing these interactions is extremely small and yet I may expect the sentence to be understood. The reason for such easy fit is called meaning. Meanings operate on a more abstract level of reality than those myriad interactions. It is on that approximate level that we say reader and writer to have the same culture. The question, then, is whether culture refers to the sum of those myriad causations, and merely offers a macro-perspective on them, or instead refers to a separate, autonomous realm where meaning follows its own laws. In sum, what are anthropologists looking for when distilling recurring meanings across widely different cultural practices? What is the status of the episteme underlying all forms of scientific knowledge which Foucault hoped to detect.[25] What did Lévi-Strauss mean by stating that the mind is structured like the world?[10] The shocking hypothesis is that the meaning system forms our thoughts in interaction with neural circuits. Meaning is a form of matter. The established academia, including social scientists, will denounce such violation of the segregation between nature and culture, between Dilthey's causality versus intentionality respectively. Anti-segregationists such as Latour for their part will denounce my taking these two concepts seriously instead of as mere constructs, mistaken purifications of the only reality: hybrids.[26] I claim that the stew has ingredients. All these denouncers had better asked themselves why it is two, and always

18

the same two, nature and culture, that have been opposed if both are not real entities. Let's consider more closely the explanatory potential of my thesis that 'culture' refers to a causality of its own, unfathomable to those studying 'nature' alone.

4. Infinity and Contingency

Social scientists, and dualists in the mind/matter debate, assume that something non-physical exists. Few of them have paused to realize how hard a claim this is. Think twice, though, before leaping to the other side in the debate between dualists and materialists. If materialists are right, that mind equals matter, then they can only claim this demystifies the mind if they ignore what it could mean for matter. What amazing thing is matter if it equals this fleeting, meaningful moment we call consciousness? The bottom-line is that we reconsider what matter means. This brings us, as is increasingly understood in consciousness studies, to quantum theory. Many sense that the solution to the two great mysteries, quantum mechanics and consciousness, may be to equate them. The first step is to realize like Henry that everything is observation and thus to shift from a materialist to a mentalist model of the universe.[27] The next step is to accept that observations could be things — meaning is matter — and to therefore develop a theory which deconstructs the distinction, allowing for the one currency of meanings interacting with feelings, without denying the duality called for by our meaning-making capacity.

The denial of a leap between synaptic connections in the brain and the actual thoughts people entertain can be found in biological models of 'neural Darwinism',[28] neural synchrony,[22] synchronous firing in the brain's electromagnetic field,[29] or competition between neural probabilities.[14] Newton's physical monism attracts. It works for the observable world. Yet, quantum theory introduced indirect methods of observation, revealing a physical pluralism in which things are not intrinsically linked to their properties, so that the particles making up matter can be matter or wave depending on the observer, and energies can exist without mass.[30] Following Stapp's interpretation,[31] quantum mechanics establishes a duality in nature between continuous associations (with probabilities) and discrete moments of conscious selection.

The famous quantum puzzle is that conscious measurement by an observer abruptly changes a particle's range of possible trajectories. When light photons are fired across a double-split screen they create an interference pattern, from which can be derived that these particles have no fixed location. They exist as probability waves; they pass through the two slits at once. However, whenever a particle is detected (measured) it chooses one slit; the interference pattern ends there. At measurement of the particle's position the wave function (described by the deterministic Schrödinger equation) appears to 'collapse' and a single actuality survives with greatly reduced future possibilities. Classical physics tells how objects affect the observer (light on the retina) but not how the latter's consciousness might influence the object so that the particle 'knows' to have been observed. Bohr's Copenhagen Interpretation stipulates, without further explanation, that observation is the reason for actualization. A recent variant proposes that the particle 'decoheres' due to its inevitable interaction with, and thus 'observation' by, other particles (whether or not these belong to a human observer). However, the unexplained then shifts to the wave function, namely the parallel existence of mutually exclusive possibilities, and to why one particle's position came out rather than another.[32] To this day no consensus exists.

The nagging problem, I submit, is caused by the mechanistic model which reasons in terms of influence. The same problem appears if we assume consciousness to emerge from influences between cells. There is no sense of a whole to which the parts might be relating. We act as if the hand rather than the person writes a letter — to paraphrase Descombes's critique of materialism.[33] One competing interpretation to Bohr's could occasion the transition from a mechanistic to a holistic approach; from influence between parts producing a whole to selection of parts from a whole (which should lead us to the cultural selection marking the human species). This competing interpretation is Hugh Everett's many-worlds view. It proposes that the wave function never collapses. Every possibility is realized in some superrealm and in the observer's corresponding 'many minds'.[34] The inflation which may repel, I regard as an advantage. The many worlds are worthy of Leibniz's famous dictum requiring from cosmologists to explain why anything exists rather than nothing. What the 'many minds' do conflict with, however, is our sense of there being one mind, one actual past for each self. Contrary to Everett, a good anthropologist will not exclude the possibility that mind requires something more than matter. That openness is achieved if we combine Everett's solution with the other no-collapse view, that of David Bohm, who claims that particles and waves lead a parallel existence (and interact). Although perfectly accounting for the data, this dual existence without deeper connection disenchanted Bohm himself.[32] I argue that human consciousness dramatically solves his problem, for it establishes a bridge between particle and wave: mind is the consistent selection of a particle from the wave. Via Bohm's dualism, Everett's many-worlds view thus permits us to shift the problem of quantum measurement from a mysterious influence to a selection among many worlds. Fundamentally, we are dealing with two worlds of matter: the natural one of probabilities and the cultural one of determined selections. The latter selects the observable universe. One particle's position at the time is destined for that universe.

Why do I say 'destined'? Here comes what I consider the main advantage of my proposal. Such predetermined totality, of which all matter partakes, strongly reminds of Einstein's universe as it is interpreted today. Of this spacetime 'loaf', as lucidly pictured by Brian Greene, observers cut very different slices of events, so that depending on position and (near-to-light) speed of observers a distant future for the one can be a distant past for the other — at the same moment.[32] In other words, physicists have agreed (against Einstein) that the universe is non-local. The observable universe, that is. The implication is staggering, at least as far as the matter constituting our thoughts (the content of our mind) goes. Everything has already happened. Now caution is in order. It is the content of what humans observe, think and so on, that has already happened. Beyond this content which I equate with Einstein's spacetime, we cannot think. It is the (cultural) selection which humans evolved (naturally selected) to observe. Hence, I do not assume a change from wave into actualized event, nor an influence of the observer in that change. I propose that the particles themselves with their indeterminate, wave-like existence belong to another world, which we cannot observe and only indirectly know from experiments (such as those famously suggested by Bell and validated by Alain Aspect). In that unobservable world nothing is predetermined, destined if you will, as opposed to the spacetime we can perceive. Particles have an open-ended position without knowledge of the future. But when we observe them, their position appears to be fixed, even if this position is a purely contingent choice to be made in the future. This bizarre condition is known as quantum entanglement. Based on Wheeler's idea to entangle the spin of particles, it has been shown that not only will a photon acquire a definitive spin value by the act of measuring, its formerly entangled but now distant companion will always choose that 'contingent' value too as if they were influencing each other instantly across spacetime (Ref. 32, p. 115). The dominant interpretation, expressed by Greene, has been to read this result as refuting the Einstein, Podolsky, Rosen argument pro locality. I argue instead that if we consider the role of consciousness in scientific observations, Einstein's special relativity turns out not to be jeopardized: the entangled photons are not influencing each other (faster than the speed of light). Anticipating on my proposal: the only thing faster than the speed of light is what has already happened. The mystery of quantum entanglement is how the contingent choice of one particle can always be the same as that of its very distant twin. My proposal is that they are not influencing each other but that the universe corresponding to their contingent choice is the one the human brain has evolved to be conscious of. Hence, the outcome we register is 'contingent' in a special way. It is a predetermined contingency (which is revealed in the entangled particle). But wasn't there supposed to be freedom? Yes, in the real world, the multiverse, where our world is one next to many others. But this multiversal reality we cannot perceive, unless through its effects in our world. Consciousness is possible thanks to a predetermined system of meanings filling it. As demonstrated by recent interdisciplinary work on visual perception,[35,36] it is great enough magic to perceive something external to the organism, so we need not be surprised that this perception is a partial selection.

Slightly wiser we may now tackle the quantum puzzle again. Detection transforms the (probabilistic) wave of a photon into a (determined) particle. Double-slit experiments have been arranged so as to postpone detection until the end of a photon's path. The experiments consistently reveal that the photons correctly decide in advance to become a particle as if these would know beforehand that the random detector would be switched on when they are reaching it at the end of their path. The way I explain such apparent premonition is that we are not dealing with influence (actualization). The wave does not decohere into a particle after detection. Rather, when the experiment shifts to the observable world, it shifts to a world predetermined and thus populated by definite particles. Here rule Einstein's laws and his god playing no dice. Counter-intuitively, this is not the 'real' world Einstein was after. It is the world of our thoughts. If we locate ourselves where our mind is, then we are standing, as it were, at the edge of this universe we observe. Everything has already happened in it. This is the universe of meanings, of the stuff we can know: laws that are eternal. It is the spacetime in which past, present and future cannot be distinguished, and of which Einstein supposed nothing beyond to exist — leading many, including himself, to wonder why it has no place for the 'now' which humans experience. His suggestion was, of course, that our everyday human experience is illusory. My hypothesis inverts his suggestion: the now is the real thing, in the more encompassing perspective. Now comes the equation of quantum mechanics and relativity theory, which is no way short of an attempt to reconciliate Aristotle's scheme of potential/act with Plato's eternal ideas: our decisions are free (quantum), but to exist in the realm of consciousness (Einstein's loaf) these and all the other related decisions, thus everything else, must have been made. Still puzzling? Think of a collective source on which individual consciousness draws, like the air we breathe. Rather than mechanistically building up consciousness from 'neural correlates', I look for the water we as fish swim in and hardly ever notice (to paraphrase Kluckhohn's famous metaphor of culture).

How come we lively recall events, without life itself being stored in the neurons? Neurons are no more than cues to reach this second environment, the collective memory-bank surrounding us. Imagine the mechanistic absurdity of our brain having to store the infinite details of every possible event encountered! It is significant that our ancestors intuitively conceived of out-of-the-blue contingency as an arrival of something already existing and interconnected with the rest of spacetime. The word 'contingency' comes from the Latin *contingere*, 'to arrive' (see also the French for 'it happened': *c'est arrivé*, cf. Ref. 37). Events arrive from somewhere: 'fate'. Significantly as well, the Sukuma farmers and herders in northwest Tanzania coin fate as *wilelo*, eagerly translated by Western missionaries as God because of the causal role which Sukuma attribute to it. The literal meaning, though, is infinity (in space and time). It so happens that the mathematical equation of (microscopic) quantum mechanics and (macroscopic) relativity theory yields this result: infinity. Mathematicians take this result as meaningless and as a proof that the two perspectives of matter cannot be reconciled. Instead, I propose that infinity, defined as fate or 'animate contingency', perfectly captures the peripheral place where the mind (world of quanta) is located in relation to its content, the universe selected (world of relativity). The world which evolution determined we can be consciousness of, differs from the 'real' multiverse out there. In regard to the particle's premonition, then, it is logical that no future contingency could fool our observations. Contrary to what moderns assume about contingency, chance and dice falling, the choice by our consciousness-and-universe has been made. There would be no (cultural) selection otherwise. We can only perceive the particle that is detected. Reminding of Everett's option, we could never detect a particle of another world, or we would be winding up there with a corresponding mind.

Surprisingly, natural selection supports this view. Our mind consistently selects an option that partakes of the totality of decisions made beforehand for the brain and that mind to exist. The spacetime we possess — we are possessed by — is the selection that permits consciousness. The quantum entanglement and non-locality we observe are logical consequences of this one spacetime forming human consciousness (neurally specified in individual states of consciousness). Rather than far-fetched I believe it almost tautological to state that whatever universe we come up with must be delimited by our consciousness. There is another way of making this statement sound more acceptable: since meaning arises only by virtue of parts relating to a 'whole', as semiotics taught us, the mind needs this enclosed totality for chunks of spacetime — which we call events — to have meaning for us. The 'now' closes the loop by which events relate to the whole and make sense. As I have argued, the part-to-whole reference serves the survival function of standing intense emotion — our escaping into the whole that is consciousness. 'What has been good for our species to think with?' is how an anthropic quantology approaches the meaning of matter. It tells what the universe we can be conscious of is like. Natural selection gave rise to the brain and to the extra access generating the mind's categories. We can thus better grasp the ongoing struggle of physicists with our naturally selected categories, such as temporality and locality, causality and materialism. That is why anthropologists could help in imagining what other categories of mind may have been selected away.

We belong to another universe than the one we observe. That multiverse we belong to is known as the quantum world: indeterminate, full of possibilities and freedom, just as our will. Since no cultural selection has determined the multiverse, we can only be sure that it differs from the observable world, and is bound to embarrass materialists. Materialism will reject physical dualism on the basis that consciousness differs not from any other phenomenon observed in nature (Thoughts would resemble each other in the way flowers do). But consciousness does differ. It is the only natural phenomenon that we directly apprehend. Not coincidently, this directly apprehended reality — the thought — does not become part of observed content. We 'observe' a thought only by deserting it and creating the next one. Again physical dualism, in this case between the content of our mind and the broader world the mind belongs to, is the idea to truly fathom.

5. Self-Organization as Free Destiny

Obviously, I do not pretend to dispose of conclusive evidence for my take on mind and matter. My first intention has been to undermine a few Western certainties. Once we have seen reductionist approaches to consciousness end up in conflations of realms and hence contradictions, we might be more willing to learn from non-Western views. The ritual initiation into Chwezi possession has besides the mechanistic method a holistic objective. It is interesting for our interdisciplinary, intercultural endeavour to discover how the holistic approach translates in mechanistic terms to produce the same meaning of spirit mediumship.

Among Sukuma farmers and herders, the everyday mode of consciousness seems generally marked by a pulsation of two opposite tendencies: a sense of 'blessing' (*lubango*) received from the ancestral spirits, and a sense of what they call 'collection' (*lukumo*) referring to accumulated goods, alliances, offspring, provisions and so on, which they earned through investment of effort. The one is subtly related to the other, so that people believe in free will, but not unboundedly so, with the happy consequence that individual responsibility is felt as shared with clan and ancestors. In mediumship, a rather uncommon practice anywhere, one side is emphasized: blessing. The subject opens up; 'subjects' itself to the spirit, but depends entirely on the latter's agency. As the Chwezi sing at night when longing for possession: "The great cooling snake emerges when the moon appears. It instils pains in my back. Maybe it has accepted." The spirit announces itself as a 'great snake' (*liyoka*) through convulsions starting from the spine. The medium depends on the snake's goodwill and on the moon cycle. The initiation to become a medium, then, introduces the other side, that of earning blessing, which constitutes the paradox: if no ritual can command the spirit, how can possession be initiated? Well, quite simply, this paradox is the point of mediumship, the novices learn. They have to accept that what they are doing, and what accounts for a consciousness of consciousness deeper than that of the non-initiated, is that they are 'stalking the stalker' (*ukusuutila lusuutila*). The spirit whom they are after is itself stalking them. Two separate systems passionate enough about each other can meet. The passion announces the pending event. What the ritual does is to take away the novice's inhibition so that, as another Chwezi song goes, "[o]n the road lies nothing but fate." Feeling in synchrony with what intrudes: how else could something be instigated that cannot be commanded?

A marked tension can be discerned with modern thought. Self-proclaimed moderns pride themselves in giving up the certainty of old (mostly religious) truths. They accept 'to live with contingency'.[38] Do the Sukuma not? Their beliefs fundamentally differ from the claims of truths sanctioned by science and by religion (which due to the intensity of their claims are logical adversaries). Their beliefs are the very embodiments of enacting contingency. As farmers responded to my queries at separate occasions: "We believe in ancestral spirits but I have never seen one." Does divination work? "We compare oracles of several diviners but one can never be sure." Indeed, premonitions can come true. Activities are deployed on the basis that divination works *and* works not. Life's ambivalence is lived, not denied, and for this purpose the Sukuma worked out practices with different experiential meanings: magic and ritual, bewitchment discourse, divination and healing, spirit mediumship. I do not have the space here to unfold the whole argument but it should be clear that the Sukuma, and other people coping daily with matters of precarious life and contingent death, not only live with contingency like moderns but — unlike moderns — accept the opposite of contingency (animate, spiritually guided chance) as a possibility too and thus live with the paradox which moderns have preferred to cast away as irrational: humans 'will' their fate. How can one be free to fulfil one's destiny? And when not fulfilling it, was this not rather one's destiny? Modern thought dislikes anything smacking of circular reasoning (or the snake biting its back) as in the Chwezi stalking the stalker. Moderns cut the snake into two — the umbilical tie if you will — separating the human into, on the one hand, a creature with free will, as in daily life, art and authorship (the status the scientist expects), and on the other hand a creature determined by other forces, as in science and religion. Segregating, instead of having the snake bite itself, indicates our trouble in understanding the paradoxes of quantum measurement and non-local space, and in taking 'infinity' for an answer. It is no coincidence that many biologists and sociologists are converging today in their struggle to figure out what self-organization (or autopoietic system) really means, one generation after the biologists Varela and Maturana and the sociologist Luhmann introduced it.[22,12]

The Chwezi idiom presents the self, the mind, as a creation of one's spirits. It recasts the holistic model of self-organization in mechanistic terms that no longer conceal the spectacular implications. The quantum theory combining Everett and Bohm distinguishes many worlds (including the one we can be conscious of) from a selection of properties (laws or meanings). It introduces the physical duality of a thing and its law. The order of laws is not four-dimensional like things, since constants are timeless and imperceptible. Subtracted of these two dimensions (time and depth) the order of laws is a two-dimensional space. My hypothesis, then, is that our universe arises from our consciousness being energized by the constants of 2D space. In other words, consciousness 'parasites' on the physical output from this order of meanings which connects with our neural apparatus of feelings. Just as we began to accept that the laws are products of our mind, I reverse the terms in the equation. Our mind is a product of the laws. Physics, sociology, and so on, develop different versions; they are all still the product of human consciousness. Conservation of energy and compensation ethics, for instance, might be energized by the same mind constant. Another such constant may be fractals. A snail's shell re-enters the 2D ratio of its spiral — the so-called golden ratio of 1.6 — into the 4D matter of which it is built. No point in getting all mystical about this: nature can, because time has selected away when nature couldn't. The amazing thing, though, is that in thought our organism can at once, like nature does over time, split off the abstract pattern from matter and re-enter it where needed, as a meaning felt in a particular way. The snail's genes feel the shell is beautiful in the sense of having no choice but to build it in that spiralling way. Humans add the meaning of beauty in their thinking so that they can consider this fractal style and choose to apply it (see Ref. 39, on fractal patterns in African textile and architecture). A neighbouring group may choose not to. How come the human mind can split off the law from the thing? Quite simply, those laws refer to a basic version that physically exists to fill our thoughts. We have unconsciously grown receptive to the laws selecting our natural environment, long after our perceptual apparatus learned to detect the natural environment's emissions. For those deeming my hypothesis exotic, I point out that I do no more than apply the fact learned from perceiving an object, namely that we become conscious of something if we are energized by that something. Is it far-fetched to surmise that this goes as well for the meaning in our perceptions? In sum, what other reason than such physical influence would there be to take our thoughts seriously?

6. Conclusion: Physical Dualism

Following Latour's plea for a symmetrical anthropology, one mediating rather than cutting apart the hybrid nature-cultures that populate our world, I have suggested an anthropic quantology.[26] Physical dualism seeks to terminate the segregation of nature and culture, without falling into the other trap, that of colonizing culture by nature (as sociobiologists did) or *vice versa* (as culturalists did). Physics too no longer defines matter in classical terms, as irrespective of the human observer, that is, as irrespective of mind. Rather than asking 'what is matter?' the question now has interdisciplinary tenor: what matter can humans be conscious of? And what have they by chance gained access to?

Of the uncountable inventions this planet has witnessed, the one of meaning — an invention that emerged under extreme conditions of natural selection — has stood the test of time simply because the invention was a discovery, of another causality than the natural environment. These meanings, or laws, rework our perception and determine consciousness. They account for the limited relevance of natural selection in determining our thoughts. Although navigating away from Darwinism, physical dualism ends up equally far from Intelligent Design. Nature's randomness fully reigns in the totality of many worlds that constitute reality. No point in hallowing the golden ratio for surfacing from natural selection. The misfits live outside the one world we can be conscious of. Any ratio is golden somewhere.

Physical dualism boldly equates quantum mechanics and relativity theory in a specific concept of contingency, which redefines infinity and combines Everett's and Bohm's interpretation of the quantum measurement problem. Rather than isolating one or the other question, it tentatively offers an explanation for the main issues at once, in the domain of both mind and matter: why we are conscious (spacetime is our sea of consciousness), why the universe we perceive is non-local and time-asymmetric (because so is the content of our mind), why we measure particles instead of waves (because a wave covers many worlds; it exists not in the spacetime selection of consciousness). Meaning requires a part-to-whole reference, just as neural cues need an enveloping sea of consciousness or memory bank. All decisions in a universe are completed before any thought can arise in it. This leads us to the anthropic principle every cosmology should integrate according to Barrow and Tipler:[40] how could matter have all the right properties to allow for planets, life and human consciousness? My answer is that we only observe our selection and easily forget about the many (invisible) worlds. Another issue that could be looked into is why we observe so little of reality, with 95% being dark energy or dark matter. Could these be traces of the new selections, universes, that sprang from our universe, with their increasing number explaining the push for our timespace to expand? Back to mind, could Libet's famous 0.5 second interval between neural processing and actual consciousness point to intervention from the second environment to turn brain into thought?

Finally, the duality of a cultural selection reigning the mind and a natural selection reigning the many worlds out there renders the amendment on Darwinian theory by Gabora and Aerts crucial:[41] the effects of natural selection on our universe can only be meaningfully accounted for if we include for every event its potentialities, rather than limiting ourselves to its one actuality in this world. In my reading of self-organization, every whole (universe) is, at another level, a part. In the neurobiological case, the part-to-whole subordination by which neurons give rise to a mind (a self) does not happen by itself, as suggested by mechanistic hopes about big enough permutations. We are enveloped by a synchronizing environment, which exists next to many other selections that do not synchronize. As long as Darwinian accounts of successful adaptations overlook this, their work will paradoxically be grist to the mill of Intelligent Design supporters wondering why there can be good solutions at all. As far as I can see, physical dualism offers the most comprehensive account, if admittedly as challenging as interdisciplinary research itself.

Acknowledgments

The research was financed by the Fund for Scientific Research in Flanders (FWO) and the Special Research Fund (BOF) of K.U. Leuven.

References

1. E. Gellner, *Reason and Culture*. London: Blackwell (1992).
2. A. Appadurai, *Après le colonialisme: Les Conséquences Culturelles de la Globalisation*. Paris: Payot (1998).
3. H. Moore and T. Sanders (Eds.), *Magical Interpretations, Material Realities: Modernity, Witchcraft, and the Occult in Postcolonial Africa*. New York: Routledge (2001).
4. G. Spivak, *In Other Worlds: Essays in Cultural Politics*. New York: Methuen (1987).
5. C. Brace, *'Race' is a Four-Letter Word*. New York: Oxford University Press (2005).
6. R. Rorty, The brain as hardware, culture as software. *Inquiry* **47**, 219–235 (2004).
7. D. Chalmers, *The Conscious Mind: In Search of a Fundamental Theory*. New York: Oxford University Press (1996).
8. M. Bloch and D. Sperber, Kinship and Evolved Psychological Dispositions:

the mother's brother controversy reconsidered. *Current Anthropology* **43** (5), 723–748 (2002).

9. T. Deacon, *The Symbolic Species: the co-evolution of language and the human brain*. New York: Norton (1997).

10. C. Lévi-Strauss, *Les Structures Elémentaires de la Parenté*. Paris: PUF (1949).

11. K. Stroeken, Stalking the stalker: a Chwezi initiation into spirit possession and experiential structure. *Journal of the Royal Anthropological Institute* **12** (4), 785–802 (2006).

12. N. Luhmann, *Ecological Communication*. Cambridge: Polity Press (1989).

13. A. Damasio, *The Feeling of What Happens: Body and Emotion in the Making of Consciousness*. San Diego: Harcourt (1999).

14. C. Koch, *The Quest for Consciousness: A Neurobiological Approach*. Englewood, Co.: Roberts & co (2004).

15. E. Baum, *What is Thought?* Cambridge, MA: MIT Press (2004).

16. D. Servan-Schreiber, *Guérir*. Paris: Laffont (2003).

17. C. Finlayson, *Neanderthals and Modern Humans: An Ecological and Evolutionary Perspective*. Cambridge: Cambridge University Press (2004).

18. A. Mosca, A Review Essay on Antonio Damasio's The Feeling of What Happens. *PSYCHE* **6** (10), `http://psyche.cs.monash.edu.au/book_reviews/damasio/index.html` (2000).

19. D. Lewis-Williams, *The Mind in the Cave: Consciousness and the Origins of Art*. New York: Thames & Hudson (2002).

20. F. Crick and C. Koch, Why neuroscience may be able to explain consciousness. *Scientific American*, special issue: the hidden mind, 94–95 (2002).

21. S. Strogatz, *Sync: the Emerging Science of Spontaneous Order*. New York: Penguin (2003).

22. F. Varela and E. Thompson, Neural synchrony and the unity of mind: A neurophenomenological perspective. In: A. Cleeremans (Ed.), *The Unity of Consciousness: Binding, Integration, and Dissociation*. Oxford: Oxford University Press (2003).

23. S. Hameroff and R. Penrose, Conscious events as orchestrated spacetime selections. *Journal of Consciousness Studies* **3** (1), 36–53 (1996).

24. A. Lutz et al., Long-term meditators self-induce high-amplitude gamma synchrony during mental practice. *The Proceedings of the National Academy of Sciences USA* **101** (46), 16369–16373 (2004).

25. M. Foucault, *The Order of Things: An Archaeology of the Human Sciences*. London: Tavistock (1970).

26. B. Latour, *We Have Never Been Modern*. London: Prentice Hall (1993).

27. R. Henry, The mental universe. *Nature* **436**, 29 (2005).

28. G. Edelman and G. Tononi, *A Universe of Consciousness*. New York: Basic Books (2000).

29. J. McFadden, *Quantum Evolution: Life in the Multiverse*. London: HarperCollins (2000).

30. J. Gribbin, *Schrödinger's Kitten and the Search for Reality*. London: Weidenfeld & Nicolson (1995).

31. H. Stapp, Quantum Interactive Dualism: An Alternative to Materialism. *Journal of Consciousness Studies* **12** (11), 43–58 (2005).

32. B. Greene, *The Fabric of the Cosmos.* New York: Penguin (2005).

33. V. Descombes, *The Mind's Provisions: A Critique of Cognitivism.* Princeton, NJ: Princeton University Press (2001).

34. J. Schwartz, *The Mind & The Brain: Neuroplasticity and the Power of Mental Force.* New York: HarperCollins (2002).

35. J. Gray, *Consciousness: Creeping Up on the Hard Problem.* Oxford: Oxford University Press (2004).

36. A. Noe, *Action in Perception.* Cambridge, MA: MIT Press (2005).

37. K. Stroeken, In Search of the Real: the healing contingency of Sukuma divination. In: M. Winkelman and Ph. Peek (Eds.), *Divination and Healing: Potent Vision*, Tucson: University of Arizona Press, 29–54 (2004).

38. U. Beck, A. Giddens and S. Lash, *Reflexive Modernization: Politics, Tradition and Aesthetics in the Modern Social Order.* Cambridge: Polity Press (1994).

39. R. Eglash, *African Fractals: Modern Computing and Indigenous Design.* New Brunswick: Rutgers University Press (1999).

40. J. Barrow and F. Tipler, *The Anthropic Cosmological Principle.* Oxford: Oxford University Press (1986).

41. L. Gabora and D. Aerts, Evolution as context-driven actualization of potential. *Interdisciplinary Science Reviews* **30**, 69–88 (2005).

THE RELEVANCE OF A NON-COLONIAL VIEW ON SCIENCE AND KNOWLEDGE FOR AN OPEN PERSPECTIVE ON THE WORLD

RIK PINXTEN

Department of Comparative Sciences of Culture
University of Gent, Belgium
E-mail: Hendrik.Pinxten@UGent.be

The so-called 'Methodenstreit' was wrongheaded, because both camps were guilty of the colonial attitude. They did not take into account the perspectives on knowledge and other traditions in any genuine way, let alone allow for conceptual openness. Without becoming exotic or unscientific we should be able to do that. The conceptual and epistemological problems involved as well as the methodological ones (praxiology, performance theory) are scrutinized.

1. The Colonial Attitude

For decades positivists and phenomenologists have been fighting each other, each claiming a superior view on science in particular or on knowledge in general. In the social sciences, and certainly in anthropology, two separate approaches have been known for over a century.[1]

The so-called social science or positivistic approach claims that objectifying is an intrinsic characteristic of science: that is to say, looking at phenomena in a deliberately detached way and thus conceiving them as an object, as foreign from one's intuitions and emotions. Any familiarity between the object of study and the subject of the researcher is considered to be a disturbing factor for proper research. Suppe, who is referred to as the best source on what became known as the received view on the positivistic interpretation of scientific knowledge, notes that the attacks on this approach by Kuhn, Feyerabend and others were rebuked in the seventies.[2] The adapted version of the received view according to Suppe, puts less emphasis on objectification and more on the internal coherence of the theory. That is to say, the theory about the object under study should construct knowledge about the object in a logically coherent way. Although the emphasis is different, it is obvious to me that this is just another way of objectification. That is to say, in order to ascribe logical coherence to a theory the phenomenon under study has to be stripped of any and all phenomenological features, and this is done most efficiently by reducing to the maximum the subjective aspects both of the phenomenon and of the process of studying it.

In the Suppe volume the famous quantum physicist David Bohm points to the weakness of such an approach. He claims that science involves communication in an equally fundamental way, specifying that communication and perception in science form a unity.[3] Hence, each unity of communication + perception stands to some extent on its own and can not readily and easily be pitched against the next one. The whole notion of (in)commensurability then becomes a rather hazy one, since theory x and theory y deal with different realities to a large extent, and can not simply be considered to be different views on the same reality. One step beyond this Bohm states that mathematical formalisms (and Suppe's logical coherence criterion in general, I would add) can not be said to be true or false in themselves, but should be qualified as relevant for x or y instead. Granted that Bohm's theory is not met with general acclaim, I think his view can not easily be discarded either. It is certainly in line with a longstanding minority position in rhetorics: Perelman and Olbrechts-Tyteca already proposed that all knowledge propositions (except for theological utterances, but they fall in a different category altogether) can be seen as subject to argumentation and persuasion, and that mathematical formalisms are not an exception to this rule.[4] Hence, communication and argumentation are of the essence in the scientific process according to this theory. I, for one, plead guilty of adhering to this school of thought.[5]

A general and rather systematically overlooked feature of the positivistic approach is what I would call its inherent colonial attitude. This becomes clear when focusing on the social sciences and the humanities. Indeed, the object of study in these disciplines is a subject and her actions and products. In the research process this subject is not readily treated as an object: people lie, hide, evade, and do many other things in order not to be treated as a mere object. In my view you should say that, in the context of research, one best starts from the view that research is being done in a double bias situation. That is to say, granted that the researcher (as a human being and hence as a social, emotional and cultural being) has biases at the conscious and at the subconscious level about the subject under study, I plead that the subject is biased as well. Malinowski's field diaries, which were published posthumously, illuminate this issue quite nicely.[6] The diaries caused quite a storm in anthropology, because Malinowski kept them in the strict sense of the term to note down his feelings and longings at any moment in the field. Hence, one finds less flattering statements in them (the great ethnographer longing for sex, or complaining about the smell of his informants, and so on). The purpose of the diary, as Malinowski intended it rightly, is to keep yourself in check over time: one can go back after the field and check the notes one made against the information of the diary about one's mindset, illness, desires of the moment and so on. In that sense, the diary is a great instrument to help and correct memory. It is, of course, not meant to be published. However, its publication brought in the open that scientists are biased. In a further step I claimed that ethnographic records show abundantly that the subjects of study hold biases too: they mistrust the ethnographer, or project expectations on him or her, or like or dislike him, and so on. Hence, the characterization of the field situation as a double biased one.[7] With the studies on experimentator-induced bias and default in so-called experimental psychology this view on social sciences and humanities can be generalised.[8] The important point here is that the colonial attitude of the positivistic researcher forbids the alteration of these conditions: with this view on science the researcher is the sole observer, the only interpreter and the final decision maker on issues of relevance, veracity and truth. If the characterisation of double biasedness holds water, then this situation should be altered altogether for the purpose of more scientific validity. Put differently, I claim we should not leave it up to openness or clairvoyance or even goodwill of the researcher to become conscious of the biases and to try and deal with them. Rather, the double biased predicament should be appreciated as a challenge which delineates clear constraints of the one-directional approach of the positivist.

The phenomenological view on social sciences and the humanities takes a different perspective on matters. Here the researcher appreciates that a certain degree of familiarity exists between the subject and herself, because she recognizes being a subject as well. The effect of this act of consciousness, however, did not lead to a break away from the one-directional approach. Rather, the leaders of this school of thought (especially from German origin) advocated that since the researcher is akin to the subject of research it follows that the researcher should then try to become like his subject by growing into her, by participating deeply in the foreign culture as a person and thus become one of them (through Einfuhlen, empathy, or other types of understanding). Once there, the researcher was believed to be able to study the other person or culture from the inside. The view is, of course, untenable: one cannot just erase the education one got and become the other person, and especially over the past decades it became clear how deeply programmed we are by religion, tastes and lifestyles, so that we cannot get a hold of our own biases in a systematic way, let alone dump them in order to be raised in the other culture. However, from my perspective on matters, this school of thought is equally and similarly one-directional as the positivistic one: the researcher is the one who holds all the cards in this view and she or he is engaging in a projected restructuring of himself or herself as the sole instrument of observation and interpretation. The subject of research is not seen as a partner or as an equally knowledgeable interactive human being here. In my words, this view is guilty of the colonial attitude to the same extent, although the approach differs on a second level. One critical step further, I claim that the phenomenological approach is likely to produce monsters just as much as the positivistic one because it does not seem to be able to digest the double biased predicament in the first place.

2. Overcoming the Colonial Attitude

Globalisation changes the world we live in in a variety of ways: those who have a good education and live in a wealthy part of the world can look forward to a good life, and those who live in a poor part will almost certainly be more destitute in a generation from today. The countries or regions where access to ICT is high will get advantages out of that condition and at the very least have the opportunity to raise the next generation in a perspective of hope and relative progress, whereas the populations from other areas face gloomy prospects and will see emigration of the more educated and the more entrepreneurial groups to the richer countries.[9] Wars of all kinds, driven by heightened identity movements of primarily a cultural or a religious stock, will be the fate of the poorer areas predominantly, and the protectionism by the rich regions via world trade agreements and through military interventions will aim at keeping the balance of inequality for as long as possible. At least that seems to be the expectation of international organisations such as UNDP.[10]

This all too brief sketch is not intended to pose as an analysis. It only indicates in a very succinct way the contextual constraints for the scientist of today. Whenever philosophers of science wrote about science in general, in the abstract as it were, I felt uneasy about their views. Surely, scientists are human beings and their work is contextualised, even if their aim is to reach so-called decontextualised knowledge. First of all they get insights (intuitive or otherwise) from the context they grow up in: e.g., Christian and Jewish scientists of the past have abundantly illustrated how metaphors and concepts of their religious upbringing have inspired them in their scientific work.[11] Similarly, the influences of Islam on mathematics have recently been documented.[12] Secondly, however, scientists have become by and large salaried intellectual workers. This puts them economically and socially in a different class from their predecessors (such as Newton or Laplace) who were economically more independent than most of their contemporaries. With the present widespread depreciation of science the vulnerability of the scientist as a worker has become an unavoidable aspect of science: contract workers abound, and the interests of industrial corporations and the military are becoming ever more important for universities and for individual careers.[13] In a very recent discussion on contract work within anthropology a debate is growing on the open advertisements in the American professional journals for jobs with the government, notably in the CIA services. It is obvious that doing contract research for the intelligence service, which recruits anthropologists under the title 'The work of a nation' and asks for studies on social, organizational and contextual factors affecting the functioning of political, terrorist and criminal groups, as well as societies), promises to be a few steps removed from what used to be known as free research.[14] It is indicative that such open advertisements are a new phenomenon, even a première, within the discipline. In older days the indirect sponsoring of anthropological work in East Asia led to a scandal and around the Second World War the ideological accusations against the founding father of American anthropology Franz Boas met with general disapproval of the discipline's professional organisations.

Alongside these developments it is clear that the world of today is shifting towards global dynamics. Hannerz summarizes the major scenarios which are advanced on this issue in the field. It is important to distinguish between the statement that interdependencies and dynamics can be detected to work worldwide and their effects. The latter can be global as well, or their relevance can be local in some sense or other. Hannerz distinguishes between three scenarios or theories in the contemporary scholarly literature:[15] a) some authors claim global dynamics triggers global culture (the Fukuyama idea: the whole world will become capitalist), while b) others hold that the world is reorganised in a limited number of cultural areas or civilizations which are on a clash trajectory (Huntington and others belong in this category). Hannerz claims that anthropologists offer a third scenario of effects, called the transformationalist scenario: in the field (at micro- and meso-level) it is not clear that grand uniformisation tendencies can be detected. Rather, myriad small shifts and changes, adaptations toward creolised or hybrid forms of culture and identity can be seen everywhere. Along any of the three views the context of the work of the scientist is seen to be evolving on a worldwide scale, with effects which differ more or less radically in each particular locus. So, all three scenarios agree on the basic issue, which is that the world is becoming more integrated, or, put differently, that isolation from others is less possible now than it was before. The consequence for the perspective of this contribution is that more and increasingly intensified contacts with individuals and groups from other cultural traditions form part and parcel of the living conditions of the population of the world. Isolation from contact and influence by cultural others is rapidly disappearing. It is my contention that the colonial attitude can still be upheld in such a constellation. However, it is obvious that such an attitude will be ever more costly and less efficient, since it will induce more misunderstandings, more frustration and possibly more confrontation and eventually violence.

One of the consequences of these contextual changes is that the arena for scientific research and the constraints of competition, research propositions, demands and rewards are becoming more and more diverse or culturally mixed. The American scene anno 2006 sees a forceful (re)emergence of religious groups challenging the status of scientific knowledge and threatening to substitute religious theories for science education in schools: Intelligent Design or creationism instead of biological evolutionary courses.[16] Their claims are voiced in terms of knowledge. The new zealots (in a biblical sense) challenge the status of science as sole or main provider of dependable knowledge. Schooled advocates and some researchers side with the defenders of Intelligent Design and some advocates refer to Indigenous Knowledge Systems to make their point.[17] It is crystal clear that the status of knowledge(s) and certainly that of scientific knowledge is under attack. Hence, the job opportunities for researchers are looking rather gloomy.

In a more general sense, the landscape of research is shifting to become an international arena, with competitors from a great variety of countries and cultures. In many western universities young and promising researchers are now offered a contract by Asian universities and laboratories. In the colonial attitude of the past the derogatory treatment of members from other cultural communities than that of the colonial power was accepted, in fear or in anger, by the former. However, in the present era this sort of treatment is not accepted or admitted anymore: the post colonial studies have produced a strong and critical countermovement, and the critique on western orientalism as a means to install and continue imperialist domination is quite established now.[18] It is clear to me that the colonial attitude I pointed to in the previous section is becoming more and more of a hindrance.

The recent history of anthropology is indicative, I think. Over the past three decades anthropology has conducted a thorough and systematic critique of its own premises, concepts and methods of the past.[19] This self-critical or reflexive anthropology has had deep impact on the discipline: other methods and other perspectives on the subject matter of anthropology are being tried out. I go into some detail here, since this particular development in one discipline may be inspiring for other sciences in the future.

3. Anthropology Looks in its Mirror

A metaphor for anthropological knowledge of the old days (the Second World War and before) was that of the mirror for man: anthropology was presumed to be able to hold up a mirror wherein a scientific theory about humanity would be visible. With the decolonisation of the world there have appeared serious cracks in the mirror: we are much less confident that our yesterday theories are still valid, and we have come to a debate now where the culture-ladenness of our theories and concepts is a most important issue.[20,21] Fundamental notions like culture, kinship,[22] religion,[23] education, value and art are rethought in a post-, less or non-colonial perspective. Questions abound such as: "is there a Chinese, an African, an Islamic, a Native American, view of the body, of the mind, of human beings or of nature that is equally valid as the western one?" and "is the western view nothing more than that (and hence not a potentially universal view)?" are coming to the fore. The universality of scientific concepts and theories is challenged, as is that of normative proposals such as Human Rights. The way to tackle these problems can not be explained in a few paragraphs, but their impact can be pointed at here.

In the 50s and 60s of the past century the metaphor of anthropology as the mirror for man was fashionable. This optimistic view proved naive and self-deluding in the decades that followed: the mirror reflected what the viewer projected on it, and not reality as such. We became deeply conscious of our prejudices and of the way the mirror image was distorted

38

by them. One step beyond this consciousness is the appeal to develop a comparative consciousness: as a social scientist it is important to study any subject in his or her context.[24] The Cherokee do not exist, except in the context of their neighbouring groups in the 18th century wide open lands of North America, or in the context of subordination vis-à-vis a capitalist WASP majority two centuries further on. The muslim is not an entity as such, but should be situated in particular contexts: pre- or post-colonial times, in national and international power relationships, and so on.

Secondly, the anthropological discipline critically looked at its concepts and its methods of the past and concluded by and large that at the very least the basic biases could be detected and that both object constructions and methods should be developed which would enable us to overcome these biases. A deep understanding of the 80s was that anthropologists are fundamentally dependent on the quality of empirical data (from field work) and that this quality depends heavily on the type of interaction one is able to set up and develop with the interlocutors in the field. In epistemological terms, social sciences are not registering or measuring first ad foremost, but develop knowledge from intensive and engaging interaction between the so-called object of study (the subjects, the informants, or whatever name we give them) and the researcher. In a sense, one might say that the research is a form of interaction, and the quality of research is dependent to a large extent on the quality of the interactions one can engage in. It goes without saying that the unidirectional view of the past, whether of a positivistic or of a phenomenological vintage, is barring from qualitative interaction, since one of the parties involved in the exchange is not really taken seriously. The only knowledgeable partner in the interaction, that is the only one who decides on the criteria for veracity, interpretation and so on, is the researcher. Ironically enough, that is the partner who has the lesser knowledge of the subject being investigated. In anthropological terms: the Indigenous Knowledge of the informant is consequently treated as of secondary relevance to the scientific knowledge about the indigenous knowledge. This entailed a discussion about the status of Indigenous Knowledge vis-à-vis scientific knowledge. I am convinced we did really conclude that discussion, for one reason mainly: both our understanding of scientific and of lay knowledge (i.c., Indigenous Knowledge) is undergoing serious changes these days. I adhere to Crombie's theory which distinguishes between at least five styles of scientific thinking, bringing some styles closer to everyday thought than any other theory of science before him.[25] For example, taxonomic thinking is demonstrably close to and sometimes identical with classificatory thought, whereas probabilistic explanation is, in its formalism, quite removed from day to day thinking.[26] At the same time, the study of counting systems, spatial reasoning and other ethnomathematical phenomena showed us that so-called native systems of thought use abstract notions, inference rules and other cognitive tools westerners have been reserving to western originated scientific thought.[27] In the wake of these new understandings the methodologies are now changing: a rise of qualitative methodological books and journals is complementing the traditional quantitative literature. Some path-breaking books have shown the way towards alternative approaches, with a high degree of relevance and hence of feasibility. I mention two of them in anthropology, since they are becoming landmarks in the discipline in my estimate.[28-30]

Bourdieu analyses the positions of the positivistic and the phenomenological school and concludes that both are unidirectional in nature and hence do not grant a position of equal partners in interpretation and decision in the research process to the subjects under study. Nevertheless, the data as well as the quality of the data from empirical research, and hence the validity of the models and theories emerging from scientific research, are dependent on the quality of what the informants are willing to offer. His critique is seconded by a constructive proposal: he invites social scientists to adopt a praxiological approach to empirical research. Basically, he sees a dialectical process in empirical study: the researcher perceives, observes, hears data which are interiorized and interpreted in the mindset of the scientist. Immediately the researcher exteriorizes her perceptions and interpretations and invites the informant's reaction on the provisional understanding about her or him. This bi-directional process goes on until both parties agree that a sufficiently valid and relevant image has been reached, representing the fallible but qualitatively grounded knowledge about the subject under study.

Hymes speaks about a performative approach to empirical study in the field, and Fabian elaborates this perspective in a systematic way. The general idea about this approach is that any objectification should be avoided, yielding a co-production or bi-directional going along with the informant in her cultural doings and sayings rather than positioning oneself as an outsider to the process one witnesses. In a general way the understanding of what is under study is gained by doing and experiencing the phenomena one is interested in through deep participation in the processes of learning, repeating or rehearsing, discussing and otherwise busying oneself with cultural skills and performances. The anthropologist or social scientist integrates herself in the processes going on and invites as much guidance and input as one can get and is allowed to get.

Decidedly, both these approaches pose a lot of questions. For one thing, they dethrone the scientific researcher in such a drastic way that reticence is the least that could be expected from the scholarly world; all of a sudden the scientist who has to gain knowledge through deep and trusting interaction proves to be involved as a subject and looses all presupposed superiority or power in the endeavour. Furthermore, the presumed safety of the old methodological criteria (repetitiveness, experimentality, communicability, and so on) does not seem to hold in any clear or direct way anymore. Rather, such difficult notions as relevance, trust, durability and the like are coming to the centre of attention. In and of itself this does not pose problems, of course. However, science as a human activity involves positioning, status and acclaim by the representative bodies of scientific quality control: the academies, the scientific boards and the like. A shift of this nature and amplitude causes unrest and invites traditional attitudes and feverish reactions: Campbell's important critique on the implicit anthropology of so-called scientific (experimental, testing, etc.) psychology of the end of the 20th century triggered extremely conservative reactions from a vast group of psychologists, and the counter-reaction of a status seeking minority who claim the status of natural science for the discipline.[31] Power and business interests are certainly at play in this case. In anthropology the hermeneutic turn seems to have won the battle so far, and hard-boiled positivists are rather the exception than the rule. Nevertheless, the discussion continues and no agreement of any deep nature seems to be reached so far: almost on a monthly basis one of the major journals in anthropology reopens the debate by focusing on society versus culture or on the question of the inclusion or exclusion of any of the traditional pillars in the discipline (language, organism, culture, history).

In the frame of this contribution and whatever the future outcome of developments, I claim it is indicative that anthropology is struggling with ways and means to escape from the colonial attitude. It is likely that it will be vulnerable for a time in doing so. On the other hand, when anthropologists do reach insights of the kind elaborated by Bourdieu, Fabian, Hymes and their likes, they will almost certainly have relevance for the other social scientific disciplines in that they showed how to come to deal with human beings in their diversity, rather than reducing the differences away in a self-controlled distancing perspective on humanity. In that sense, the seedbed of anthropological discussion could even prove to be relevant for science in general: other intuitions, other interpretative frames, and other metaphors will be reappraised and allowed to function alongside the western ones on the future. In my view I compare this possible shift to the one known as the conscious appraisal of the phenomenal illusion. When the geocentrists of the Renaissance kept saying that the sun revolved around the earth because that is what was in the Bible and because that is what we can all plainly see with our own eyes, they fell for this illusion: what you see corresponds to what is out there in reality. The heliocentrists rightly understood that what they saw was an illusion, and corrected that wrong image by thinking and observing in combination. In my view something similar is at stake here: the illusionists believe that what they see and think is obviously superior to what others have brought forward throughout the centuries. The former think that any and all items of their vintage are hence beyond doubt. The shift which may be triggered by a better understanding of the situatedness and context-dependency of all knowledge will, in my appreciation, also be that of a move away from a phenomenal illusion. We have to become conscious of contextual and other constraints of our knowledge, such that we learn to interpret and evaluate models and theories within a different mindset than that of the naive realist. The proposals on praxis and on performance may help the social scientist to adopt this more warranted perspective on knowledge by actually learning the skills in the interaction of research. Obviously, a lot of work needs to be done in order to equip such a perspective on research with the methodological sophistication we were used to in the older schools of thought. But at the very least a start has been taken with the works by the pioneers mentioned.

References

1. E. Service, *A Century of Controversy*. New York: Random House (1989).
2. F. Suppe, *The Structure of Scientific Theories*. Chicago: University of Illinois Press (1977).
3. D. Bohm, Science as Perception–Communication. In: F. Suppe (Ed.), *The Structure of Scientific Theories*. Chicago: University of Illinois Press, 374–391 (1977).
4. C. Perelman and L. Olbrechts-Tyteca, *La Nouvelle Rhétorique*. Paris: PUF (1958).
5. R. Pinxten, Ethnomathematics and its Practice. *For the Learning of Mathematics* **14** (2), 23–25 (1994).
6. B. Malinowski, *A Diary in the Strict Sense of the Term*. London: Routledge (1968).
7. R. Pinxten, *When the Day Breaks*. Frankfurt: Lang Verlag (1970).

8. D.T. Campbell and F. Stanley, *Quasi-experimental design*. Washington D.C.: Rand McNally (1969).

9. M. Castells, *Conversations with Manuel Castells*. Oxford: Blackwell (2002).

10. S. Fukuda, *Cultural Liberty in Today's Diverse World*. New York: UNDP Report (2004).

11. G. Hadamard, *De l'invention*. Paris: Albin Michel (1960).

12. G.G. Joseph, Foundations of Eurocentrism in Mathematics. In: A. Powell and M. Frankenstein (Eds.), *Ethnomathematics*. New York: SUNY Press, 51–60 (1997).

13. N. Chomsky, *The Humanities In Cold War America*. New York: Free Press (1999).

14. H. Price, America the ambivalent. *Anthropology Today* **21** (5), 1–2 (2005).

15. U.Hannerz, Macro Scenario's. *Social Anthropology* **5**, 17–35 (2003).

16. C. Tourmey, Intelligent Design for Anthropologists. *Anthropology News* **46** (9), 5 (2005).

17. A.J. Petto, Nature and Design. *Anthropology News* **46** (9), 4 (2005).

18. E. Said, *Culture and imperialism*. New York: Basic Books (1990).

19. J. Fabian, *Time and the Other*. New York: Columbia University Press (1984).

20. M. Owusu, The Usefullness of the Useless: Ethnography of Africa. *American Anthropologist* **80**, 310–334 (1978).

21. C. Geertz, *Available Light: Anthropological Reflections on Philosophical Topics*. Princeton: Princeton University Press (2000).

22. D. Schneider, *American Kinship*. Chicago: Chicago University Press (1989).

23. L. Orye, *Verborgen hypotheses in menswetenschap*. Brussel: VUB Press (2001).

24. L. Nader, Comparative consciousness. In: J. Borofsky (Ed.), *Assessing Cultural Anthropology*. New York: Wiley (1993).

25. A.C. Crombie, *Styles of Scientific Thinking in the European Tradition*. London: Duckworth (1994).

26. S. Atran, *Cognitive foundations of natural history*. Cambridge: Cambridge University Press (1990).

27. M. Ascher, *Ethnomathematics*. Pacific Grov: CA Books (1991).

28. P. Bourdieu, *Practical Reason*. Cambridge: Polity Press (1981).

29. D. Hymes, *In Vain I Tried to Tell You*. Philadelphia: University of Pennsylvania Press (1981).

30. J. Fabian, *Power and Performance*. Berkeley: California University Press (1990).

31. D.T. Campbell, On the conflicts between biological and social evolution and between psychology and moral tradition. *American Psychologist* **30**, 1103–1126 (1975).

AN ATLAS FOR THE SOCIAL WORLD: WHAT SHOULD IT (NOT) LOOK LIKE? INTERDISCIPLINARITY AND PLURALISM IN THE SOCIAL SCIENCES

JEROEN VAN BOUWEL

Centre for Logic and Philosophy of Science
Ghent University, Belgium
E-mail: Jeroen.VanBouwel@UGent.be

Starting from the analogy between theories and maps, I will spell out which interdisciplinary approach to the social sciences can provide us with the atlas we need to navigate in the social world. After comparing the features of theories and maps in Sec. 2, I elaborate how different social theories can collaborate or get into a dialogue in Sec. 3, summarizing the different strategies that have been defended for interdisciplinarity in social science: theory-, method-, metaphysics-, and question-driven interdisciplinarity, which I will illustrate with actual proposals made by, *inter alia*, World-Systems Analysis, Critical Realism and Economics Imperialism. Building on the framework of explanatory pluralism I have been developing before, I will make a case for question-driven interdisciplinarity in Sec. 4. My argument for question-driven interdisciplinarity will be illustrated in Sec. 5 by discussing recent developments in economics (i.e., the debate between the orthodoxy and heterodox theories, the pleas for pluralism, and the impact of globalisation — and related institutional developments — on economics as a discipline). In conclusion, the contours of an adequate atlas for the social world should become clearer; when to use the different maps, how to activate the dialogue between social scientific disciplines in order to draw the different maps, and the risks of globalisation for social science (and adequate map making).

1. Introduction

When one tries to develop a scientific view on the social world, she will most likely consult the social sciences. But, a first introduction to these sciences will soon end in despair: the plurality of social theories, forms of explanations, social categories, etc., is endless. In this chapter, I would like to explore how we can compare, contrast, promote dialogue, etc. between social theories, between social worldviews. Therefore, different forms of interdisciplinarity will be discussed and evaluated in the quest for the best atlas of the social world. Let me start with explaining the appropriateness of the metaphor of the atlas.

2. The Social Sciences, the Puzzle and the Maps

Elaborating an adequate theory of the (social) universe has often been understood as solving a puzzle. The different pieces found in different scientific disciplines should eventually be put together in order to get the jigsaw science has been looking for. Recently, however, several philosophers of science have been claiming that the metaphor of the puzzle should be replaced by the metaphor of the map (e.g., Refs. 1–4). In what follows, I will develop the analogy between scientific theories and maps, and show what it can teach us concerning interdisciplinarity in the social sciences. I will distinguish some features of maps and then relate them to scientific theories.

Imagine you have a day off and you want to spend it in Brussels. While preparing your city-trip, you wonder which map you will have to bring, and you wander off into the features of maps:

M1 *Drawing a map.* There is no single correct way or method of map-making. The projection, the colours, the scale of a map can vary depending on the goal, function or interest of a map. However, maps have to live up to a minimum of social conventions (in order to be more than lines on paper); without these conventions we would not be able to produce and use the map.

T1 *Elaborating scientific theories.* A lot of philosophy of science has been driven by the idea of making the (one correct) scientific method that will bring us good scientific theories, explicit. Whether one such method exists, is doubtful given the variety of methods applied on a multitude of research topics by scientists pursuing good scientific theories. However, they all live up to minimal social conventions; without these we would not understand scientific theories.

M2 *Goals, functions and interests of maps.* Different types of maps serve different functions and interests. A map of the Brussels' subway system will be less apt if one wants to embark on a walk passing the museums. Depending on the goals and interests, one can design a variety of maps of one city: a map of the subway tracks and stations, a map providing all the street names, a map with tourist attractions, museums, the best restaurants, the shopping malls, etc.

T2 *Goals, functions and interests of theories.* Every scientist is focussing on a specific aspect of the world, a specific problem. This has not only resulted in different scientific disciplines, but in very divergent, multiple perspectives within one discipline; the study of economic phenomena, for instance, is being done from a variety of perspectives, cf. *institutional, behavioural, feminist ...* economics.

M3 *The accuracy of maps.* Maps can be more or less accurate. For instance, the map of Brussels drawn by Lodovico Guicciardini in 1581 is less accurate than most contemporary maps.[a] No map is completely accurate.

T3 *The accuracy of theories.* Theories can be more or less accurate. Ptolemic physics are generally considered to be less accurate than Newtonian physics.

M4 *The use of maps.* Whether a map is good or bad, does not only depend on its accuracy, but as well on how adequately it serves the goals and interests for which it has been drawn or is to be used. A less adequate map can give us a clearer view, while a more detailed map can be useless or too hard to use given our goals.

T4 *The use of theories.* Whether a theory is good or bad depends (among other things) on the handiness, the easiness to apply or use it. In the sciences we often encounter calculations or predictions based on less accurate theories, that, however, are more handy to be applied (e.g. employing Newtonian physics instead of relativity theory for calculations, or economic theories based on an extreme *homo economicus* hypothesis instead of a bounded rationality hypothesis or a broad *homo sapiens* hypothesis).

M5 *A complete map or the ideal atlas.* There is no map or atlas which is exhaustive or complete; we can always add extra information or more details. It is not plausible that a city map could be drawn which would serve all our goals or interests (we can possible have). Our goals and interests can always change.

T5 *A theory about everything.* It is not plausible that there is a the-

---

[a]Talking about historical maps, I would like to mention that there are many examples to be found of (inaccurate) maps used as propaganda tool, e.g. the map of Mercator of the British Isles, where Scotland is just as big as England. Hence, accuracy sometimes has to make room for 'higher' goals. Even today maps do give away ideological and other preferences of the producer and user, e.g. in choosing to have Europe, the USA or Australia in the centre of a world map.

ory that can serve all our goals and interests, even though some scientists keep on dreaming about it (cf. Ref. 5). Theories select some elements of the world, and do exclude others; they are always partial (e.g. illustrated in economics by Ref. 6). Hence, an ideal, everlasting theory seems to be impossible, taking also into account that our interests and goals can change.

M6 *The art of making maps evolves.* Not only our goals and interests might change, but also the techniques and conventions applied in making maps are adjusted to the circumstances of their time.

T6 *Theory-making evolves.* As is clearly shown by the history of science, making theories, scientific research, scientific techniques and conventions, etc. do change and adapt themselves to the circumstances and possibilities of their time. Old theories are being replaced by new ones, new fields of research are being explored, etc.

M7 *Map and world.* We can experience the correctness or truth of a map in our relation to the world, the city of Brussels in this case. A map is not merely a construct (that has to live up to social conventions and has to be interpreted), it is as well the map of something, a region, a city. As Giere puts it: "Nevertheless, maps do manage to correspond in various ways with the real world. Their representational powers can be attested by anyone who has used a map when traveling in unfamiliar territory." (Ref. 1, p. 25)

T7 *Theory and world.* The correctness or truth of a theory is tested (by our relation) with reality. This is a pretty complex procedure, and has to deal with a series of philosophical problems, *inter alia,* underdetermination (e.g. in economics, see Refs. 7, 8). The map metaphor shows that having more than one good theory does not necessarily imply that theories are merely social constructs, without any relation between the representation and the represented.

Selecting the best city map thus involves a lot of pondering. So does drawing the best possible atlas of the social world. In what follows, I will discuss some attempts to compile an atlas, to reduce the variety of maps and theories of the social world; ways to compare, contrast, or organize a dialogue between social scientific disciplines and theories.

3. Interdisciplinary Strategies: Collaboration and Division of Labour between the Social Scientific Disciplines

Elaborating an atlas of the social world has often involved conflicts, indifference or collaboration between the different social scientific disciplines, like economics, sociology, anthropology, political science, etc. In what follows, I will summarize and exemplify the different strategies that have been defended for the collaboration and division of labour between the disciplines in social science. Many proposals have been launched which has resulted in an extensive list of disciplinary terms: interdisciplinarity, multidisciplinarity, transdisciplinarity, unidisciplinarity, postdisciplinarity, non-disciplinarity, mono-disciplinarity, etc. I will, however, distinguish four strategies, i.e., (3.1) imperialism, (3.2) insulation, (3.3) synthesis, and (3.4) the question-driven strategy.

3.1. *Imperialism or Colonisation*

The debate about the division of labour between the different disciplines is often characterized in terms of power balances: imperialism, colonialism, integration, isolation, etc. The first strategy we can distinguish is sometimes called imperialism or colonisation, i.e., one discipline (or the dominant method or theory of a discipline) wants to unify the social sciences by seizing and incorporating the other disciplines. Economics imperialism based on neo-classical economics, is a good example.

The central idea of economics imperialism, or the economics 'takeover', is that economists want to improve knowledge in the social sciences outside economics through an economics takeover, applying the dominant theory and method in economics far beyond its original home.[b] Mäki describes it as follows:[11] "Economics imperialism is a matter of persistent pursuit to increase the degree of unification provided by rational choice theory by way of applying it to new types of explanandum phenomena that are located in territories that are occupied by disciplines other than economics."

It involves a theory- and method-driven extension of economics to topics that go beyond the traditional issues addressed by economists including the theories of consumer choice, production, markets and macroeconomics; all social behaviour can be explained by using the tools of economics. For

---

[b]The label *imperialism* might sound too strong, but Gary Becker — author of the *locus classicus*[9] of economics imperialism — actually agrees with the label: "This definition of 'economic imperialism' is probably a good description of what I do." (Ref. 10, p. 39).

instance in the political sciences, there is an increasing amount of articles published from this economics imperialism perspective applying rational choice theory (cf. Refs. 12, 13).[c] Whether this improved theories and explanations in these disciplines is controversial, as will be discussed in Sec. 4.

Another example in this strategy of imperialism and colonisation is sociological imperialism, as noticed by, e.g. Sayer:

> "But sociological imperialism is not much better, indeed it can lead to strikingly similar problems. It encourages academics to emphasize not what is relevant and important for understanding social phenomena but whatever promises to raise the profile or educational capital of their discipline. [...] This is easy to see in others, harder for disciplinarians to see it in their own behaviour. Faced with any attractive theoretical innovation, we should always ask — is it attractive because it seems to enlarge the claims of the discipline or because it's a good explanation of the phenomenon concerned. [...] When sociologists say science has to be understood as a social construction, does this appeal because it's a better explanation of science, or because it advances sociological imperialism — others don't realise it's a social construction, but sociologists know it is and they understand that better than anyone." (Ref. 17, p. 1)

3.2. *Insulation*

A second possible strategy concerning the division of labour between the social sciences (in order to obtain a good atlas of the social world) is insulation, i.e., no collaboration between disciplines. There are social scientists, e.g., historians as Evans and Windschuttle,[18,19] that want to defend their field against intruders, and that are not reaching out for enrichment by other disciplines. This kind of monodisciplinary research is restricted to one research discipline (and to one branch or specialization within a re-

---

[c]Economics imperialism is not limited to political science. Sociology, anthropology, history, etc. are being confronted with it as well. Gary Becker has applied economic theory to non-eonomic (non-market) problems like criminality and family planning, problems that are usually studied by sociologists. In sociology, James Coleman has contributed a lot to the foundation of sociology on a rational choice model.[14] Contemporary follow-ups on this project can be found in Ref. 15. In anthropology the model has been applied by e.g., Ladislav Holy,[16] Burton Pasternak en Samuel Popkin. The impact and success do however seem to be limited. Applications of the rational choice model in history can mainly be found in the work of the Cliometrics, history-writing based on neoclassical economics.

search field). People working within one discipline study the same research objects, share the same paradigm, use common methodologies, and speak the same 'language'. That is to be defended against intruders, according to this strategy.

3.3. *Collaboration Leading to a New Synthesis*

A third possible strategy does not want to unify the social sciences by incorporating all of them into one discipline, but rather proposes that the boundaries between the disciplines be transcended by a comprehensive framework, an articulated conceptual or metaphysical framework leading to an overarching synthesis. This might be called *unidisciplinarity* or *transdisciplinarity*. The legitimacy and meaning of the separate disciplines evaporates (in that sense *postdisciplinarity* or *non-disciplinarity* could be adequate terms as well). Let me present two examples: Critical Realism and World Systems Analysis.

The Critical Realist perspective was born out of a vigorous critique on the positivist conception of science. It pleads for the reorientation of social science, unveiling the *epistemic fallacy* committed by positivists. The *positivist* social scientist analyses statements about being solely in terms of statements about knowledge, and thus reduce ontology to epistemology. Therefore, as a reaction to this neglect, it is "opportune to develop a perspective on the way that social reality is" (Ref. 20, p. 154).

Hence, after the unveiling of the epistemic fallacy, the focus should be replaced on ontology. Central in the focus on social ontology, then, figures the transcendental argument for social structures, elaborated by Roy Bhaskar. He derives an account of a metaphysics of science by enquiring what the world must be like before it is investigated by science, and for scientific activities to be possible. Bhaskar's transcendental realism defends the existence of social structures and society as follows:

> "... conscious human activity, consists in work on *given* objects and cannot be conceived as occurring in their absence. A moment's reflection shows why this must be so. For all activity presupposes the prior existence of social forms. Thus consider *saying, making* and *doing* as characteristic modalities of human agency. People cannot communicate except by utilizing existing media, produce except by applying themselves to materials which are already formed, or act save in some other context. Speech requires language; making materials; actions conditions; agency resources; activity rules.

> Even spontaneity has as its necessary condition the pre-existence
> of a social form with (or by means of) which the spontaneous act
> is performed. Thus if [as previously argued] the social cannot be
> reduced to (and is not the product of) the individual, it is equally
> clear that society is a necessary condition for any intentional human
> act at all." (Ref. 21, p. 34)

This argument is used to establish that "the social cannot be reduced to
(and is not the product of) the individual, it is equally clear that society is
a necessary condition for any intentional human act at all" (Ref. 21, p. 34).
Bhaskar's ontological framework (including, e.g., the *Transformation Model
of Social Activity*) becomes a distinctive feature of the Critical Realist contributions to social science. Using a common (unified) ontological view of
social reality across the different social sciences, the Critical Realist contributions promise the unification of the social sciences, cf., Christopher
Lloyd:

> "If their practice were to be based on the realist-relational ap-
> proach it would provide a framework for simultaneously explaining
> particular acts, events, patterns of behaviour, consciousness, and
> structural change. [...] It is because of the deeper relation of partly
> intentional behaviour to both the given structural conditions of be-
> haviour and the production, reproduction, and transformation of
> structures, that action-oriented and structure-oriented history can
> be united on a more fundamental level. Such a unified science
> would ideally then incorporate all the existing empirical and theo-
> retical social and historical studies." (Ref. 22, p. 195)

The growing amount of social scientific literature employing Critical Realism, e.g., the work of Margaret Archer in sociology, Tony Lawson in economics, Heikki Patomäki in political science, etc., might strengthen Critical
Realists in their conviction that social science moves towards unification and
its optimal state.

A second example of the synthesis strategy comes from World System
Analysis, more specifically from its founder Immanuel Wallerstein. He
has been defending a unidisciplinarity in the social sciences that might be
reached by imposing a unifying concept, i.e., Wallerstein's historical system.
It would synthesize the social sciences into an historical social science.[23]

3.4. *The Question-Driven Strategy*

A fourth strategy is not so much theory-, method- or metaphysics-driven, but propagates collaboration between social scientific disciplines based on questions. This can be understood as *interdisciplinarity* in which several disciplines collaborate (problem- or question-driven); in which concepts, methodologies, or epistemologies are explicitly exchanged and integrated, resulting in a mutual enrichment. The result of this collaboration is something more than any of the separate disciplines could have offered. The disciplinary structures are however being respected.

The question-driven approach might as well be *multidisciplinary*. This means a collaboration in which several disciplines juxtapose their results (concerning one problem or question), adding breadth and available knowledge, information, and methods. The disciplines speak as separate voices, in encyclopaedic alignment. There is no systematic attempt at integration, combination or crossdisciplinary work (no integration of concepts, epistemologies, or methodologies. The degree of integration between disciplines is restricted to the linking of research results).

As I will defend in the next paragraph, the question-driven approach might provide us the best possible atlas of the social world, even though critics do not stop repeating that both interdisciplinary and — definitely — multidisciplinary endeavours usually have the effect of reinforcing existing disciplinary and departmental boundaries.

4. Defending Question-Driven Interdisciplinarity for Drawing the Atlas

In Sec. 2, I have elaborated the contemporary view on the map-like features of scientific theories. Let us now evaluate the different strategies presented in Sec. 3 in the light of that map metaphor and the quest for the atlas of the social world. As I will refer to different forms of explanations provided by different social theories, I will first introduce a framework for understanding explanations in the social sciences in paragraph 4.1. Then, I will go on to the actual evaluation of strategies in paragraph 4.2, and in paragraph 4.3, I will elaborate a defence for question-driven interdisciplinarity.

4.1. *Providing Good Explanations*

Dealing with the plurality of theories and explanations in the social sciences, we need some philosophical instruments to *make the explananda as explicit*

52

as possible, and pay attention to the *underlying epistemic interests*. The erotetic model of explanation seems a good candidate to help us; it regards explanations as answers to why-questions. It can be used to show how different questions about one social fact can lead to the use of different forms of explanations and different social theories.[24,25] We will present a framework for explanations that will help us to get a grip on the plurality of explanations.

Analyzing social scientific practice, different explanatory requests can be distinguished. We do not consider the explanatory requests and motivations mentioned here as the only possible ones, but we do believe they are omnipresent in social science practice. We distinguish at least four types of explanatory questions:

(plain fact) Why does object a have property P? Was the fact that object a has property P the predictable consequence of some other properties of object a?

(P-contrast) Why does object a have property P, rather than the (ideal) property P'?[d]

(O-contrast) Why does object a have property P, while object b has property P'?

(T-contrast) Why does object a have property P at time t, but property P' at time t'?

Firstly, we can have non-contrastive, explanation-seeking questions concerning plain facts, e.g. of the form: Why does object a have property P? Was the fact that object a has property P the predictable consequence of some other properties of object a? These non-contrastive questions can have different motivations. One possible motivation is sheer intellectual curiosity (the desire to know how the fact 'fits into the causal structure of the world', to know how the fact was produced from given antecedents via spatio-temporally continuous processes). A more pragmatic motivation is the desire to have information that enables us to predict whether and in which circumstances similar events will occur in the future (or the anticipation of actions of persons/groups). Another possible motivation concerns causally connecting object a having property P to events we are more familiar with (Form: "Is the fact that object a has property P causally connected with events we are more familiar with?").

The form these explanations of *plain facts* (answers to non-contrastive questions) have, can be to show how the observed fact was actually caused,

---

[d] P and P' are supposed to be mutually exclusive.

which implies providing the detailed mediating mechanisms in a non-interrupted causal chain across time, ending with the explanandum, or — considering the second motivation — the explanation can follow the covering law model.

Secondly, explanation-seeking questions can require the explanation of a contrast, e.g. of the form: Why does object a have property P, rather than property P'? (P-contrast); Why does object a have property P, while object b has property P'? (O-contrast); Why does object a have property P at time t, but property P' at time t'? (T-contrast). The explanations of contrasts can have a therapeutic function, or are motivated by 'unexpectedness'. They isolate causes that help us to reach the ideal (P-contrast), comparing the actual fact with the one we would like to be the case. Or they help us to remove the observed difference, or they could be meant to tell us why things have been otherwise than we expected them to be (T- and O-contrast — they both have the form: Why does object a have property P, rather than the *expected* property P'?).

The form of a *contrastive* explanation enables us to obtain information about the features that differentiate the actual causal history from its (un)actualized alternative, by isolating the causes that make the difference; this information does not include information that would also have applied to the causal histories of alternative facts.

By making the different possible explanatory requests explicit (cf. O-, P-, T-contrasts or a plain fact) the motivation and the explanatory information required will be taken into account. It can be shown that one social fact can be the subject of different questions, and hence of different forms of explanation. Consequently, taking into account the explanatory question is not something of secondary importance, as it decides on which form of explanation will be used. To be able to answer these different kinds of explanatory questions in the best way possible we will need different forms of explanation.

Summarizing, an explanation is an answer that should be evaluated in relation to a question that is a specific request for information (and the precise meaning of the question is therefore important). Making the questions as explicit as possible will show that a(n apparently) similar question about one social phenomenon may, given that interest and contexts select distinct objects of explanation, result in very different answers/explanations as explanatory purposes might be different; explanatory questions require answers in which the most *adequate* explanatory information is provided (not just any *accurate* explanatory information — cf. *supra* on the accurate and adequate maps). This framework for understanding the plurality of explanations provided by social theories, will help us in the evaluation of the interdisciplinary strategies (and clarifies the *question*-driven strategy).

4.2. *Interdisciplinary Strategies Evaluated*

The first strategy, i.e., (a) imperialism, has received a lot of critique, both from within economics, i.e., from the so-called heterodox economists, as from the 'colonized' disciplines, e.g., political science. The critiques of the heterodox economists will be discussed in Sec. 5. Let us mention some of the critiques formulated in political science.

Critics state that the growing body of formal rational choice theory literature is irrelevant. According to them, this literature is not answering important societal questions, 'real-world problems'. E.g., Stephen Walt:

> "In this sense, much of the recent formal work in security studies reflects the 'cult of irrelevance' that pervades much of contemporary social science. Instead of using their expertise to address important real-world problems, academics often focus on narrow and trivial problems that may impress their colleagues but are of little practical value. If formal theory were to dominate security studies as it has other areas of political science, much of the scholarship in the field would likely be produced by people with impressive technical skills but little or no substantive knowledge of history, politics, or strategy." (Ref. 13, p. 46)

Therefore, the relevance of academic political science for (parts of) the public is decreasing, according to the critics. Robert D. Putnam warns us for the danger of *policy research* migrating towards *schools of public administration*, just as happened with practical economic studies that changed the economics department for *business schools*. "If one compares the size of economics departments and business schools in today's academy, the cost of reducing a social science to sterile theoretical endeavors is obvious." (Quoted on **www.paecon.net**)

The increasing 'colonization' of political science by economists has led to so much discontent that an anti-imperialist movement, *Mr. Perestroika*, was created.[e] The Perestroikans' main focus is that major journals of the field have been preoccupied with publishing research that conforms to the

---

[e]The first steps of the movement were taken in the year 2000, with a mass e-mailing by 'Mr. Perestroika'. Schram discusses the Mr. Perestroika movement in political science:

economics takeover features. A lot of attention is paid to the institutional factors that favour economics imperialism. One striking example is the American Political Science Review:

> "One might imagine that the American Political Science Association preaches that the best governing system, despite all its faults, is a democratic one, but APSA luminaries obviously display grave doubts as to how far democracy ought to be allowed to go. From inception, the Association never entertained the wildly radical notion of conducting internal elections. What rules is a cozy arrangement whereby a committee chosen by the president nominates its successor members who picks the next governing council who pick the next president, and so on. Disgruntled political scientists link the absence of democracy in the organization to the suffocating disciplinary dominance, especially in the last decade, of formal models and rational choice theory."

Now as "the APSA of late has been run by rational choice exponents or sympathizers and its flagship journal, the American Political Science Review, reflects their unbending bent."[27] So, the economics imperialism gets an institutional push.

But let us not discuss the (very important) institutional aspects here, and stick to epistemological issues and our quest for the best atlas of the social world. Political scientists have been pointing out that the economics takeover is theory- and method-driven, not problem- or question-driven. The method dictates the problem, not *vice versa*. The focus is on the development of techniques, not on substantial political questions. Ian Shapiro has elaborated on this distinction between theory- and method-driven on the one hand, and problem- and question-driven on the other.[28] Problem- or question-driven "is understood to require specification of the problem under study in ways that are not mere artefacts of the theories and methods that are deployed to study it" (Ref. 28, p. 598). Method-driven research leads "to self-serving construction of problems, misuse of data in various ways, and related pathologies summed up in the old adage that if the only tool you have is a hammer everything around you starts to look like a nail" (Ref. 28, p. 598). For example, "making a fetish of prediction can undermine problem-driven research via wag-the-dog scenarios in which we

---

"the ways in which contemporary social science all too often fails to produce the kind of knowledge that can meaningfully inform social life." (Ref. 26, p. 836)

elect to study phenomena because they seem to admit the possibility of prediction rather than because we have independent reasons for thinking it worthwhile to study them" (Ref. 28, p. 609).

Economics imperialism is thus perceived as eager to prove that the dominant economic theory and method is superior by the self-serving construction of political problems and by substituting existing theories and/or methods in political science. Applying the map metaphor, it defends that given the successfulness of one map (which helps us to solve economic problems), we should use the same scale, colours, projection, etc. for all other possible maps. Imperialism is not so much driven to find answers to unaddressed political questions, but rather driven to get the social world into one theory, with one preferred form of explanation. This runs against the idea of explanatory pluralism (cf. *supra*) and results in a suboptimal situation as concerns the amount of explanation-seeking questions that can be dealt with.

Sociological imperialism, another version of the interdisciplinary strategy (a) imperialism, has similar defects. It is based on a denial that there is anything they need to know about on the other side of the disciplinary boundary (a boundary which allows sociologists to externalise difficult problems), e.g., critic Ian Craib points at the denial of the internal world. The realm of the 'I' — that is our capability to receive something from outside and make it our own, to make something of what we are constructed through — thus creating something different — has always created problems for sociology.[29] The sociological reductionism does preclude certain problems to be analysed and (forms of) explanations to be formulated (and as such the advantages of a plurality of maps disappear).

The pendant of imperialism is (b) insulation, the second strategy. It is another form of parochialism, of not being able to think outside the framework of your own discipline.[f] Scientific disciplines often rely on a tradition — on a canon of great examples — which can provide the resources to achieve great breakthroughs, but might as well lead to conceit. The unwillingness to open up your discipline for methods, theories and form of

---

[f]Sayer: "Disciplines are parochial; they tend to be incapable of seeing beyond the questions posed by their own discipline, which provide an all-purpose filter for everything. Where the identity and boundaries of a discipline are strongly asserted and policed, it can stifle scholarship and innovation. One of the worst manifestations of this is the production of lists of 'recognized' journals, as in economics, which are considered by the RAE to count as acceptable for publications. It would be disastrous if this were ever to happen in sociology." (Ref. 17, p. 2)

explanations from other disciplines, or to collaborate with other disciplines, guarantees that some problems and questions will not get the best possible answer. Neither will this strategy lead to the best possible atlas of the social world.

Let us now move on to the evaluation of next strategy, (c) synthesis. As an example of this strategy, I have pointed to the Critical Realist project. It proposes a way of unifying the social sciences that starts with Bhaskar's *a priori* or necessary truth concerning social ontology based on a (questionable) transcendental argument; the existence of social structures is based on a transcendental derivation as quoted above. I do not want to argue that social structures (or other ontological aspects of the Critical Realist's stance) do not exist, but that the way it has been defended by Bhaskar and Critical Realism in general, is problematic, just as it has been problematic in earlier attempts to impose preconceived ontological ideas in the (philosophy of the) social sciences.[30] The attempt to justify the claim that the world has indeed the form argued for in transcendental realism does not convince (or, better, does not convince me more than other stands in the unending battle of metaphysical intuitions we experience in the philosophy of the social sciences). Moreover, as the ontological choice made by Critical Realism does have an impact on methodological options, I want to warn for an *ontological fallacy*: taking an *a priori* ontological stance that transposes or reduces epistemological and methodological matters into an ontological matter. Analogous to the *epistemic fallacy* it points at a failure to sustain adequately the distinction between ontology and epistemology, that is, a failure to deal with both ontology and epistemology in a non-reductive way.

Whichever starting point we prefer in studying the social world, we will always adopt some ontological assumptions (it is unavoidable and necessary). With Critical Realist applications however, the ontological assumptions are 'proven' to be true *a priori*, and this raises serious doubts on whether they could at all be revised. Secondly, starting from the Critical Realist ontology has some methodological consequences that are insufficiently spelled out. The methodological consequences of the Critical Realist's ontology seem to follow 'automatically', and hence do not have to be spelled out. There is a lack of attention paid to the form of explanations and to methodology in general.[31]

If one is convinced that the relation between the individual and the structure is correctly described by (a version of) the Transformational Model of Social Activity (TMSA) — as Critical Realists do — one will not consider explanatory theories that are not in line with TMSA, e.g., rational choice theory or theories using the Covering Law model, but that might provide good (and better) answers to some explanation-seeking questions. These answers would be considered (at least) incomplete by Critical Realists. Due to Critical Realism's lack of reflection on the usefulness of different forms of explanation in the social sciences and on pragmatic aspects of explanation, good and useful explanatory information will be lost (cf. *supra*, and Refs. 24, 31). The focus is on ontology (and their convictions of how the social world *really* is), at the expense of methodology. One should be wary of the heavy metaphysical furniture imposed by Critical Realism, and of its *politics of metaphysics*. It is a good example of metaphysics-driven interdisciplinarity aiming towards synthesis.

Wallerstein's proposals go in the same direction: reframing the historical social sciences around the idea of historical systems, will limit explanatory information, e.g. by excluding possible information on an individual agency-level. Hence, the theory-driven unification of World Systems Analysis will not be able to answer to the plurality of (forms of) explanation-seeking questions, elaborated in paragraph 4.1.

It will help to invoke the metaphor of maps here, once again: just as we can design many different maps of a city depending on the goals or interests we have using them (e.g. a map of the subways, a map containing all the street names, a map pointing at the shopping zones, etc.), we can scientifically represent nature and society in different ways. As there is no objectively correct set of goals or interests, we cannot have an ideal, single and/or complete account of nature. The different social scientific theories present us different maps (some are good, others are bad, I am not defending that anything goes, that every map/theory should be accepted). Sometimes these maps concern different social phenomena, sometimes they deal with the same phenomenon. But, just as accurate maps of the same city are supposed to be consistent, these different scientific accounts of one social phenomenon can be consistent. Ending up with there being only one map available (e.g. the map of the subways) is clearly suboptimal.

Thus, unifying social science under the banner of the *a priori* Critical Realist ontology (and its methodological implications), or imposing the concept of historical system, or economics/sociological imperialism, or disciplinary insulation, does not seem the right way to overcome the intellectual division of labour in studying the social world as they cannot not take into account the plurality of knowledge-interests (and the difference these imply in the explanatory information that is required), neither the plurality of existing forms of research and explanation in the social sciences. Useful explanatory information would get lost if the social sciences were organized in their way and some explanation-seeking questions would not, or inefficiently, be addressed. Therefore, let us get back to the central problems of interdisciplinarity and the division of labour in the social sciences, and start the defence of a question-driven interdisciplinarity.

4.3. *Defending Question-Driven Interdisciplinarity*

The central problems to be solved are — given the plurality we find in the explanatory practice of social scientists: (a) to what degree should we integrate the plurality of theories, methodologies and forms of explanation; (b) what is the purpose of integration or what drives the integration? Starting with the latter question (b), the integration can be theory-, method- or question/problem-driven (cf. *supra*, Ref. 28), or, I would like to add, metaphysics-driven. A good example of a theory- and method-driven integration, as we have just discussed, is the so-called *economics imperialism*, which tries to unify the social sciences based on neo-classical economics and rational choice theory. Another theory-driven unificationist project is Wallerstein's *world-systems analysis*. Critical Realism is an example of metaphysics-driven unification. However, having pointed at the deficiencies of these approaches, I want to defend that the integration of the social sciences — bridging social knowledge — should be question- or problem-driven, aiming at efficiently answering (explanation-seeking) questions; questions that might require different kinds of explanatory information (cf., paragraph 4.1).

Answering the first question (a), then, i.e., to what degree should we integrate the social sciences, can be done by stressing the need for unification (and a form of unidisciplinarity), or by cherishing plurality. The question-driven approach opts for the second, but is, nevertheless, critical to the current disciplinary division of the social sciences. It does encourage us to cross — or institutionally dismantle — disciplinary borders and to avoid insulation, in order to find the best answer to an explanation-seeking question. Rather than driven towards unification, it is driven by a quest for the best answer, maximally using (and comparing) the plurality of theories, methodologies and forms of explanation present in social scientific practice. Hence, it cherishes plurality and claims that pluralism is optimal for the social sciences as the different forms of explanations, theories and methodologies provide us with different kinds of useful explanatory information;

depending on your motivation or knowledge-interest, one of these different kinds of explanatory information is the most apt. The parallels with maps are obvious. The different forms of explanation (explaining a fact, a P-contrast, etc.) discussed in paragraph 4.1, can only be efficiently elaborated and used within the question-driven interdisciplinarity; defending a theory-, method- or metaphysics-driven unification (like Critical Realism) would imply that some (useful) forms of explanation would be lost, and wanted kinds of explanatory information would become unavailable to us.[g] How the question-driven strategy might as well be employed as a critical tool in analysing current controversies in the social science, will be illustrated in the next section.

5. Elaborating the Atlas in Economics

In this section, I will discuss some recent developments in the discipline of economics and analyse how these developments are supported or criticised by the different strategies of interdisciplinarity we have distinguished in Sec. 3. In paragraph 5.1, I start with pointing out some important changes in the history of the discipline and relate those changes to globalisation. The reaction on economics imperialism within the discipline of economics will be discussed in paragraph 5.2, where the debate between the orthodoxy and heterodox theories is commented upon. In paragraph 5.3, we discuss the Critical Realist contribution (and its synthesis strategy) to economics. Finally, I will show in paragraph 5.4 that the map metaphor and the idea of question-driven interdisciplinarity can show us a fruitful way out of the orthodoxy–heterodoxy debate towards better economics.

5.1. *The Impact of Globalisation*

If we have a look at the history of economics, we will notice that at the time of the institutionalisation of economics in universities (in the years from the 1880s to the 1910s), a plurality of approaches existed in Western Europe and the USA. This plurality was in line with the different roles that economics played within the different nation-states.[32] Let me sketch some aspects of the international diversity in ways of doing economics.

In 19th century Germany, the interests of economists were shaped by their subordination to administrative demands — which lead to an em-

---

[g]We have defended this point in a more detailed and technical manner, in Refs. 24, 25. In these articles we make the idea of 'best answer or best explanation' more explicit.

phasis on applied policy questions with the state as a central actor in the economy. This might explain the historicism that is so characteristic for German economics: the focus should be on solving particular problems, not on 'discovering' timeless laws. In the United Kingdom, the practitioners of economics considered themselves mainly as scientists with a theoretical mission, not policy-oriented:

> "in marked contrast with the Germans, the majority of British economists at the eve of World War I were seeking to uncover abstracts laws for economic behaviour that would be independent from the larger historical context." (Ref. 32, p. 413)

Contrary to Germany, the central actor is not the state, but the individual's behaviour motivated by self-interest. In France, university economists remained by tradition juridical and literary until the 1930s, and avoided theory. The fragmentation of institutions in French economics might have contributed to the fact that:

> "France never produced a distinctively French school of economic thought with a coherence comparable to its British and German counterparts during the same period. By and large, French economics at the turn of the century conceived of itself more as a discourse on an ever-changing social reality than as a science in search for universal laws." (Ref. 32, p. 422)

A glimpse at the situation of the United States, tells us that objectivity and the scientization of economics were — contrary to the French situation — important given "their distrust of the political underpinnings of academic knowledge and their conception of social science as 'objective expertise'." (Ref. 32, p. 435) How has economics been developing ever since?

The last decades of the 20th century — in a world characterized by globalisation — we noticed a tendency for this national distinctiveness to decrease and disappear within the discipline of economics: less plurality and diversity, and more internationalisation and homogenisation (as is documented for Western Europe in a recent study by Coats).[33] As Roger Backhouse concludes:

> "American-style professional economics has spread within academic economics. [...] it is arguable that economics is now, to a much greater extent than in 1945, a single market, in which the nationality of economic ideas (and in a sense of economists) is much harder to define." (Ref. 34, p. 37–38)

This resulted in the dominance of what is called the *mainstream* or orthodox approach (cf. *infra*). The disappearance of national styles and local varieties of economics shows, for instance, in the evolution towards *one-size-fits-all* theories. This has been criticised by economic policy-makers, e.g., by Joseph Stiglitz in his critique of international organizations.[35] This very short sketch of two stages in the history of economics leaves us with the question about what the consequences of this globalisation have been for the topics related to interdisciplinarity and plurality.

5.2. *Orthodoxy and Heterodoxy in Economics*

The *mainstream* or orthodox approach (cf. a version of neoclassical economics) in economics, is the one that feeds the economics imperialism we have discussed above. We have already mentioned some of the critiques of political scientists on interdisciplinary economics imperialism. The imperialism has, however, been criticized within economics as well. The critics can be grouped under the label of heterodox economics. The heterodox economists are worried about the increase (proportionally) of journal articles — and education and research in general — using the orthodox approach. It has some pernicious consequences for the discipline, according to them.

First of all, the critics state that the orthodox approach leaves important economic or political questions unanswered, that they do not answer real-world problems.[h] Economics students state that their economics textbooks, based on neo-classical theory (using mathematical models), do not live up to their expectations. The courses are disconnected from reality. They point to their failure to answer contemporary questions, to deal with economic phenomena with which students are being confronted in their lives (the role of financial markets, pros and cons of free trade, global inequality, etc.). Nor do students learn anything about how to behave as a social and economic being:

> "Resort to mathematical formalization when it is not an instrument but rather an end in itself, leads to a true schizophrenia in

---

[h] "No reality please, we're economists" Cf. "the abandonment of anything like empirical testing of macroeconomic models" (Ref. 36, p. 44). "'Economics for economics sake' will soon become — indeed, already is — the battle-cry. At this point, someone might observe that economics is too important a subject to be left to economists. Is that why students are increasingly choosing Business Studies and Business Managements over Economics?" (Ref. 36, p. 46).

relation to the real world. Formalization makes it easy to construct exercises and to manipulate models whose significance is limited to finding 'the good result' (that is, the logical result following from the initial hypotheses) in order to be able to write 'a good paper'. This custom, under the pretence of being scientific, facilitates assessment and selection, but never responds to the questions that we are posing regarding contemporary economic debates."
(From the students' petition on www.paecon.net)

Secondly, according to the critics of the orthodoxy (and its economics imperialism), there is no discussion about methodological questions as: When are these formal methods the best route to generating good explanations? What makes these methods useful, and consequently, what are their limitations? What other methods could be used in economics? The methods dictate, the questions come second. Critics perceive a theory- and method-driven takeover focussing on substituting the heterodox contributions, not on plurality. The heterodox ideas are only listened to when framed in orthodox methods.

As a group of Cambridge PhD-students states:[37]

"only economic knowledge production that fits the mainstream approach can be good research, and therefore other modes of economic knowledge are all too easily dismissed as simply being poor, or as not being economics. Many economists therefore face a choice between using what they consider inappropriate methods to answer economic questions, or to adopt what they consider the best methods for the question at hand knowing that their work is unlikely to receive a hearing from economists."

It makes them conclude that:

"the dominance of the mainstream approach creates a social convention in the profession that only economic knowledge production that fits the mainstream approach can be good research, and therefore other modes of economic knowledge are all too easily dismissed as simply being poor, or as not being economics." (*Ibid.*)

Besides the worries about important questions being left unanswered, and methods being imposed rather than scrutinized, there is the repeated complaint about publications in top journals. Research of Hodgson and Rothman has pointed out that there is an important institutional and geographic concentration (they speak of an oligopoly) on the market of economics journals, which reinforces the orthodox approach.[38] Geoffrey Hodgson and Harry Rothman analysed which authors published in which academic top 30 journals in economics, and who the editors of those journals were. Analysing the citation impact, it was revealed that 70 % of journal editors were located in the USA. Concerning journal article authors, 65 % of them were located in U.S. institutions and twelve U.S. universities accounted for 22 % (M.I.T., Harvard, Chicago, Stanford, Princeton, Berkeley, Yale, Wisconsin, University of Michigan, Northwestern, University of Pennsylvania en Rochester).[i]

The oligopoly Hodgson and Rothman unveiled raises questions about the possibilities for change and innovation: "There are grounds to presume that the dominance of the profession by a few leading institutions is likely to reduce the diversity in approaches and beliefs." (Ref. 38, p. 180–181) "In contrast, the present domination of global research by a few institutions is likely to foster 'lock-in' where a limited number of departments defend their traditional ideas and approaches against all new-comers." (Ref. 38, p. 185) Here, again, top journals have a big influence on curricula (publish — in top journals — or perish), textbooks, promotions, the distribution of research funding, future research topics, etc., which might reinforce the mainstream's epistemic agenda.

As an alternative the heterodoxy has been developing its own publication channels since the 1970s:

> "the coalescing of economics around a central core stimulated the rise of new 'heterodox' groupings that brought together economists who felt that their ideas were being systematically excluded from the profession's main journals." (Ref. 40, p. 316)

We might wonder whether the heterodox economic schools have not turned too much on themselves, not interacting with the orthodoxy? The creation of their own journals, congresses, etc. provides an opportunity for exchange, specialization and attention for their heterodox approach, but applying this strategy might lead to getting isolated, or isolate oneself.

---

[i]A concentration confirmed by more recent research: "Economics is by far the most concentrated of the disciplines. More than 67 % of economics faculty received their doctorates from a top-ten PhD program, and almost 82 % earned doctorates from a program in the top twenty."[39] Wu also mentions that professors from top economics departments tend to dominate the pages of the most prestigious journals.

The worries about the state of economics have only recently led some of the critics to organise themselves in an 'anti-imperialist' movement, namely the Post-Autistic Economics Network (PAECON), demanding more attention for heterodox theories and protesting against the growing dominance of the orthodoxy.

> "The movement began in France in June 2000, when a group of economics students, under the banner *'autisme-économie'*, published on the web a petition protesting against:
>
> - economics' 'uncontrolled use' and treatment of mathematics as 'an end in itself', and the resulting 'autistic science',
> - the repressive domination of neoclassical theory and derivative approaches in the curriculum, and
> - the dogmatic teaching style, which leaves no place for critical and reflective thought.
>
> That debate began on the 21st of June, when the French national newspaper, *Le Monde*, reported on the students' petition and interviewed several prominent economists who voiced sympathy for the students' cause. Other newspapers and magazines followed suit." (Edward Fullbrook on the history of PAECON, `www.paecon.net`)

So did students, researchers and professors of other departments. In the Anglo-Saxon world *The Cambridge–27* (27 PhD-students in economics at Cambridge University, UK) published a similar petition called *Opening Up Economics*.[37] Another example is *The Kansas City Proposal*,[11] and a petition started up by Harvard-students and -lecturers.[42] The success of these petitions points to a growing discontent that deserves to be analysed.

After having documented the discontent about the economics imperialism of the orthodoxy within economics, we will now check the alternatives offered by the heterodox economists. As an example, I will introduce the Critical Realist alternative to orthodox economics.

5.3. *Critical Realism in Economics and its Pleas for Pluralism*

Tony Lawson is one of the most prominent defenders of Critical Realism in economics and the most developed account of Critical Realism in economics up to now can be found in his *Economics & Reality*.[20] I will not summarize his critique on mainstream economics (this can be found in Ref. 43), but just have a look at his alternative and evaluate his proposal in the light of interdisciplinarity.

A very central issue in Lawson's proposals concerning the reorientation of economics, is the unveiling of the *epistemic fallacy* (cf. *supra* in paragraph 4.2): The *mainstream* economists analyse statements about being solely in terms of statements about knowledge, and thus reduce ontology to epistemology. After the unveiling of the epistemic fallacy, the focus should be replaced on ontology, or as Steve Fleetwood puts it:

> "Clearly any such awareness [that the basic (deductivist) method of *mainstream* economics is inappropriate to its subject matter] presupposes *a prior analysis* of the nature of social phenomena — as does any project of developing an alternative. Hence, for the critical realist project, socio-economic ontology figures centrally."
> (Ref. 44, p. 129, *my italics*)

Central in the focus on (socio-economic) ontology, then, figures the transcendental argument for social structures, elaborated by Roy Bhaskar. As such, Lawson is applying the metaphysics-driven strategy we have explored above within economics. Lawson does, however, defend pluralism at the same time. How do we have to understand that?

In his 'Back to Reality' Tony Lawson pleads for more pluralism in the academy and the opening up of the economics discipline — criticizing the monism of *mainstream* economics.[45] This will imply a return to variety and greater pluralism in method, according to him. The question should be raised whether Lawson's contribution is pluralistic, as the idea behind the metaphysics-driven strategy seems to be to replace the current orthodoxy, and as such introduce a new monism.

In developing his critical realist ontology and methodology as an alternative to the mainstream economics methodology, Lawson does indeed add an alternative to the methodological options available in economics, increases diversity, and — in that sense — does contribute to more pluralism in the academy. But, some might have a more demanding idea about pluralism: one that considers pluralism as stimulating intellectual exchange and discussion, where the focus is on complementarity (and an idea of when to use the complements), rather than on an isolated or non-communicating diversity. Does Lawson's contribution live up to these demands?

Lawson's rejection of the *mainstream* mode of explanation, the Covering Law model, cannot be counted as a good start for a pluralist; a pluralist would be interested in checking the compatibility and consistency of different modes of explanation. John Davis points in the same direction, when he questions heterodoxy in economics:

> "(m)any of those who are the strongest proponents of methodological pluralism — heterodox economists — are entirely critical of neoclassical economics. Their motivation, I would conjecture, is not that their own theoretical approaches are *also* correct — a theoretical pluralist view — but rather than neoclassical economics is mistaken and misguided in its most basic assumptions, and that their own approaches remedy the deficiencies of neoclassicism — a theoretical monist view." (Ref. 46, p. 209)

So, according to Davis, the self-declared pluralists of the heterodoxy (and we can include Lawson here) do formulate alternative methodologies that are not so much meant as complements, but rather as substitutions of the *mainstream* economics. Referring to the map metaphor and the framework for explanations, we might ask whether a version of the Covering Law model and some versions of reductionism cannot be considered as complements to Lawson's proposals, so prediction and other motivations for explanations can be addressed, rather than be rejected and substituted by Lawson's alternatives.

Lawson's quest for heterodox economics is not so much focusing on elaborating compatibility with *mainstream* economics, but rather to create an own alternative, that tends to be substituting, rather than complementing. (An alternative that was established on — or wrapped in — an ontological reorientation, made hard by a transcendental argument, which led to epistemological consequences, of which we might ask whether they are products of an *ontological fallacy*.) So, Lawson presents himself as a pluralist against the *mainstream* monism, although it seems too early to conclude that Lawson is more of a monist, I want to pay attention to the risk of *strategic pluralism*. This strategic pluralism, as used by Ron Giere,[47] refers to groups that advocate pluralism as

> "primarily just a strategic move in the game of trying to dominate a field or profession. Those in the minority proclaim the virtues of pluralism in an effort to legitimate their opposition to a dominant point of view. But one can be pretty sure that, if the insurgent group were itself ever to become dominant, talk of pluralism would subside and they would become every bit as monistic as those whom they had replaced."

If Lawson and the heterodoxy in general really want more pluralism in the academy, more attention should be paid to complementarity with some *mainstream* approaches, to assure that they stay 'on speaking terms' with the orthodoxy and do not isolate themselves (or get isolated) all too much; to make sure that the best possible atlas for the economic world, with a diversity of maps, will be elaborated.

5.4. *Question-Driven Interdisciplinarity and the Map Metaphor*

Which lessons can we draw from these recent developments in economics? On the one hand we have the orthodoxy and its interdisciplinary strategy. On the other hand we have exemplified the heterodoxy with Critical Realism and its strategy of synthesis (and unification) of the social sciences. The sketch of the history of economics has shown us the transnational tendencies and globalisation of the discipline and the loss of national or local diversity. The globalisation reinforces the orthodoxy, a process that — according to critics — has been helped by the orthodox oligopoly in top journals (and related teaching curricula, promotions, etc.). The orthodoxy's institutional dominance is not necessarily a consequence of epistemic superiority. The selection of articles for publication might not (merely) be based on scientific quality, but on applying the 'right' method or theory. An emphasis on this 'right' method and theory might imply that some (important) questions about society are not addressed, as heterodox economists claim. What is the alternative to the heterodoxy?

The increasing weight and dominance of the orthodoxy limits the space available for the heterodoxy. The latter decided to organize its own journals, conferences, etc. Its plea for pluralism might be understood from this perspective, as a way to create more space for themselves. The lack of making their relation with competing approach more explicit, shows however that the pluralism is not aiming at complementarity.

The question-driven interdisciplinarity and the map metaphor show us a way out of this stalemate between orthodoxy and heterodoxy towards a more optimal state for the social sciences, emphasizing the shortcomings of both orthodoxy and heterodoxy. The dominance of the orthodoxy narrows the scope and the ability to answer questions of economics in the long run. Invoking the map metaphor, one sees that the economics imperialism will lead to maps that are all of the same scale, colours, projection, etc. This trend towards monism renders the social sciences suboptimal.

But elaborating the atlas merely on the basis of heterodox theories does not seem to be the best option either. The heterodoxy might be right in emphasising some shortcomings of the orthodoxy — their inability to answer some important questions about our societies, the focus on theoretical problems, the *self-serving construction of problems* (cf. *supra*). However, the heterodoxy does not search for complementarity, prefers isolation, loses valuable kinds of explanatory information available in the orthodoxy (cf. covering laws that help in answering explanatory requests concerning expectability).

A loss of valuable explanatory information can be avoided by adopting the question-driven interdisciplinary strategy. Different theories (both orthodox and heterodox) answer different questions raised about economic phenomena; pleading for monism will not lead to always having the best map available. As we have documented the globalising and homogenising tendencies in economics, the question-driven strategy provides us with a tool to avoid a suboptimal epistemic state of economics, to avoid insulation, and to stimulate dialogue and comparison.

6. Conclusion

Contemporary views on scientific theories have been emphasising the analogies with maps, as I have illustrated in Sec. 2. Applying the map metaphor is a critique on the view that scientists should be looking for a comprehensive scientific theory of the (social) universe, understood as solving a puzzle. The different pieces found in different scientific disciplines should then eventually be put together in order to complete the jigsaw according to this puzzle metaphor. The map metaphor, however, by emphasising the importance of interests and goals in developing scientific theories and the plurality this implies, rejects such a puzzle(d) view as leading to a epistemically suboptimal science.

In Secs. 3 and 4, I have shown the impact of the map metaphor on the debate about interdisciplinarity in the social sciences. A framework for understanding the plurality of explanations in social sciences was introduced, and helped us to search for the best option for the division of labour in the social sciences, i.e., question-driven interdisciplinarity. Making the different kinds of (explanation-seeking) questions (and the underlying interests) explicit can be linked with taking not only accurateness, but as well the adequateness of maps into account. The question-driven approach will as such result in the best possible atlas for the social world.

Clinging on to the metaphor of the puzzle would only misguide us. A look at the current state of economics from the puzzle perspective could lead an observer to conclude that the puzzle is close to being finished. However, the professional dominance and homogeneity of the orthodoxy might not prove its *epistemological* superiority. Helped by the metaphor of map making and the framework for explanatory pluralism, we can recognize the risks of globalisation for social science (and adequate map making). A question-driven approach has shown us that better economics might not be the result of blindly following the heterodox economists either; an optimal atlas for the social world is based on the plurality of non-monistic approaches (which can be orthodox and heterodox) driven by (explanation-seeking) questions which activate the dialogue between social scientific disciplines in order to draw the different maps.

Acknowledgments

The author is Postdoctoral Fellow of the Research Foundation–Flanders (FWO).

References

1. R. Giere, *Science without Laws*. Chicago: The University of Chicago Press (1999).
2. P. Kitcher, *Science, Truth, and Democracy*. Oxford: Oxford University Press (2001).
3. S. Mitchell. Why integrative pluralism? *Emergence: Complexity & Organization* **6**, 81–91 (2004).
4. M. Baghramian, *Relativism*. London: Routledge (2004).
5. S. Weinberg, *Dreams of a Final Theory*. New York: Vintage Books (1994).
6. U. Mäki, Theoretical isolation and explanatory progress: Transaction cost economics and the dynamics of dispute. *Cambridge Journal of Economics* **28**, 319–346 (2004).
7. K. Sawyer, C. Beed and H. Sankey, Underdetermination in economics: The Duhem–Quine thesis. *Economics & Philosophy* **13**, 1–23 (1997).
8. W. D. Hands, *Reflection without Rules. Economic Methodology and Contemporary Science Theory*. Cambridge: Cambridge University Press (2001).
9. G. Becker, *The Economic Approach to Human Behavior*. Chicago: University of Chicago Press (1976).
10. R. Swedberg, *Economics and Sociology. Redefining Their Boundaries: Conversations with Economists and Sociologists*. Princeton: Princeton UP (1990).
11. U. Mäki, Explanatory ecumenism and economics imperialism. *Economics & Philosophy* **18**, 237–259 (2002).

12. D. P. Green and Ian Shapiro, *Pathologies of Rational Choice Theory: a Critique of Applications in Political Science*. New Haven: Yale University Press (1994).

13. S. M. Walt, Rigor or Rigor Mortis? Rational Choice and Security Studies. *International Security* **23** (4), 5–48 (1999).

14. J. Coleman, *The foundations of social theory*. Cambridge, MA: Belknap (1990).

15. P. Hedström and R. Swedberg (Eds.), *Social Mechanisms. An Analytical Approach to Social Theory*. Cambridge: Cambridge University Press (1998).

16. L. Holy, *Anthropological Perspectives on Kinship*. London: Pluto Press (1996).

17. A. Sayer, Long Live Postdisciplinary Studies! Sociology and the curse of disciplinary parochialism/imperialism. `http://www.lancs.ac.uk/fss/sociology/papers/sayer-long-live-postdisciplinary-studies.pdf` (1999).

18. R. Evans, *In defense of history*. New York: W.W. Norton (1997).

19. K. Windschuttle, *The killing of history: how literary critics and social theorists are murdering our past*. New York: Free Press (1997).

20. T. Lawson, *Economics & Reality*. London: Routledge (1997).

21. R. Bhaskar, *The possibility of naturalism*. Brighton: Harvester Press (1979).

22. C. Lloyd, *The structures of history*. Oxford: Blackwell Publishers (1993).

23. I. Wallerstein, *Unthinking Social Science: The Limits of Nineteenth-Century Paradigms*. Cambridge: Polity Press (1991).

24. E. Weber and J. van Bouwel, Can we dispense with structural explanations of social facts? *Economics & Philosophy* **18**, 259–275 (2002).

25. J. van Bouwel and E. Weber, Remote Causes, Bad Explanations? *Journal for the Theory of Social Behaviour* **32** (4), 437–449 (2002).

26. S. Schram, Return to Politics. Perestroika and Postparadigmatic Political Science. *Political Theory* **31** (6), 835–851 (2003).

27. K. Jacobsen, Revolt in Political Science. *Post-autistic economics newsletter* **6**, article 6 (2001).

28. I. Shapiro, Problems, methods, and theories in the study of politics: Or what's wrong with political science and what to do about it. *Political Theory* **30** (4), 596–619 (2002).

29. I. Craib, *Classical social theory*. Oxford: Oxford University Press (1997).

30. J. Watkins, Methodological individualism: A reply. In: J. O'Neill (Ed.), *Modes of Individualism and Collectivism*. London: Heinemann, 179–184 (1973).

31. J. Van Bouwel, Questioning structurism as a new standard for social scientific explanations. *Graduate Journal of Social Science* **1** (2), 204–226 (2004).

32. M. Fourcade-Gourinchas, Politics, institutional structures, and the rise of economics: A comparative study. *Theory and Society* **30**, 397–447 (2001).

33. A.W. B. Coats (Ed.), *The development of economics in Western Europe since 1945*. London: Routledge (2000).

34. R. Backhouse, Economics in mid-Atlantic: British economics, 1945–95. In:

Coats (Ed.), *The development of economics in Western Europe since 1945*. London: Routledge, 20–41 (2000).

35. J. Stiglitz, *Globalization and its discontents*. New York: W.W. Norton (2002).

36. M. Blaug, Ugly Currents in Modern Economics. In: U. Mäki (Ed.), *Fact and Fiction in Economics. Models, Realism and Social Construction*. Cambridge: Cambridge University Press, 35–56 (2002).

37. The Cambridge 27, Opening Up Economics: A Proposal By Cambridge Students. *Post-autistic economics newsletter* **7**, July. `www.paecon.net` (2001).

38. G. Hodgson and H. Rothman, The editors and authors of economics journals: a case of institutional oligopoly? *The Economic Journal* **109**, 165–186 (1999).

39. S. Wu, Where Do Faculty Receive Their PhDs? *Academe* (July/August), 53–54 (2005).

40. R. Backhouse, *The Penguin History of Economics*. London: Penguin Books (2002).

41. Kansas City Proposal, `http://www.btinternet.com/~pae_news/KC.htm`

42. Harvard petition, `http://www.petitiononline.com/ec10alt/petition.html`

43. J. Van Bouwel, Explanatory pluralism in economics: Against the mainstream? *Philosophical Explorations* **7** (3), 299–315 (2004).

44. S. Fleetwood, Situating Critical Realism in Economics. In: S. Fleetwood (Ed.), *Critical Realism in Economics. Development and Debate*. London: Routledge, 127–135 (1999).

45. T. Lawson, Back to Reality. *Post-autistic economics newsletter* **6**, article 2, `http://www.btinternet.com/~pae_{n}ews/review/issue6.htm` (2001).

46. J. Davis, Comment. In: A. Salanti and E. Screpanti (Eds.), *Pluralism in Economics. New Perspectives in History and Methodology*. Cheltenham: Edward Elgar, 207–211 (1997).

47. R. Giere, Perspectival Pluralism. In: S. Kellert, H. Longino and K. Waters (Eds.), *Scientific Pluralism*. Minnesota Studies in the Philosophy of Science, 26–41 (2006).

WORLDS OF LEGITIMATE WELFARE ARRANGEMENTS: A REALISTIC UTOPIA ON PENSIONS

PATRICIA FRERICKS and ROBERT MAIER

Faculty of Economic and Social Sciences
Universität Hamburg, Germany
E-mail: patricia.frericks@uni-hamburg.de

Department of Interdisciplinary Social Sciences
Universiteit Utrecht, the Netherlands
E-mail: r.m.maier@uu.nl

Pension entitlements are based on certain assumptions. On the one hand, there are assumptions about the individual substantiated in life-course norms that focus on labour market participation. On the other hand, there are assumptions about societal necessities, such as intergenerational and intragenerational justice, economic growth, pension scheme sustainability and so forth.

The development of European welfare states led to a variety of different pension norms as a calculation principle for building up full pension entitlements. In all countries there are, first of all, increasingly more strict wage-related pension entitlements, and, varying per country, entitlements based on residency as in Denmark and the Netherlands, or pension care credits as part of the French, German and Austrian systems, and also a variety of other specific pension determining factors as well as a growing number of private pension schemes. In this article we analyse these different pension determining factors by studying different European pension reforms, taking theoretical considerations into account. The aim is to answer the question as to whether there are ideal ways of combining these factors. Our ultimate goal is to outline a legitimate and sustainable pension system.

Our 'realistic utopian approach' combines recent political, social and economic reform considerations with normative and theoretical ones and presents an original way of studying pension policies in the EU.

1. Introduction

Welfare arrangements are one of the facets that enable the reproduction of present-day polities. Reproduction points to continuity, to the necessity of social organizations to ensure their ongoing and continued existence. Pensions, the example of welfare arrangements we use in this chapter, contribute in several ways to this reproduction. First of all, pensions provide monetary resources that are necessary for purchasing basic survival goods (consumption, rent etc.). Secondly, receiving a pension marks a status passage from being an active participant in the labour force to a non-employed person, i.e., a pensioner. From the perspective of the individual, this passage can be conceived of as a transformation of social identity, and from the perspective of the employers, the employees (and their organizations) and the state, as an instrument for regulating the labour force. Thirdly, under certain conditions, the system of pensions, by attributing rights over resources to non-employed individuals (such as a mandatory retirement age, or having paid contributions etc.), guarantees a certain amount of autonomy to these non-employed people. They neither become dependent on kinship relations, nor will they be a burden on the polity. In other words, pensions achieve, in an original way, forms of intergenerational solidarity. In these three different ways, pension systems institute a modern form of welfare arrangements that helps to sustain present-day capitalist welfare states.

If, following Wallerstein,[1] we conceive of the world system as a conglomerate of polities with, on the one hand, various forms of association among them, such as the EU, and, on the other hand, of competition, it follows that the various types of polities that exist (such as nation states, states, empire states, city states, etc.) are partly independent from and partly dependent on each other. They are partly independent because they are, in principle, given their sovereignty, free to decide which social and political forms they institute. However, at the same time, because of the multiplicity of instituted and informal links between the various polities, each polity is severely constrained by these links. Decisions by one specific polity to institute social and political arrangements that subvert the existing links could seriously threaten the continuing existence of this polity.

As a next step, we introduce another important condition for our analysis, namely capitalism. Capitalism can be defined in a first approximation as a form of production of goods (of all kinds, including services) to be sold on the market at, where possible, a profit. The majority of the goods sold and bought on the market are indeed produced in a capitalist way, in addition to goods and services provided by the welfare state or by voluntary activities. Participation in capitalism is realized by adopting the roles of employers and employees, and assuming a labour market, and the roles of sellers and buyers on the various markets, again creating social identities. To be able to conclude contracts is a necessary precondition, presupposing autonomy and therefore elements of citizenship, and in particular the well-known civil and political rights. Capitalism, in its various historical guises, has never been a pure system. In addition to capitalist forms of production and markets, there have always been forms of social organizations that are not organized in accordance with the principle of capitalist production and exchange, such as the household economy for example. There are several reasons why capitalism always was and is increasingly becoming a multimodal system. Let us take education for example. The forms of capitalist production and exchange have become more technical, requiring an ever more highly qualified workforce. It is true that nowadays we are witnessing the development of a knowledge-intensive economy. Longer years of general and specialized education in the last century (an average of 1 year more for every ten years) have been instituted. The organization and the funding of the education system could not be taken over by purely capitalist forms of production and exchange, because competencies and qualifications are not goods that can be freely exchanged. People cannot sell their competencies or qualifications. It is only the use of these capabilities that can be sold. A different mode of economic functioning had to be invented, which is the state controlled one, financed through compulsory monetary transfers, such as taxes or social contributions. This economic mode grew in a significant way in the last century, in European countries in particular. Not only education, but also important parts of the healthcare system and all the other arrangements of what is called the welfare state, such as pension systems, are included in this economic mode. A wide variety of new social identities has been created within this economic mode, such as the identities of the unemployed or of pensioners.

Finally, the reproduction of present-day polities necessitates informal exchanges, achieved either in household economies or through voluntary work. The household economy, for example, involves all kinds of informal exchanges and is organized around unpaid care work, such as care for children, care for the elderly and care to sustain day by day the participants of the household. Once more, the different roles taken in the household economy create social identities, such as, particularly in the past decades, the identity of the male breadwinner, the housewife, the carer, etc.

Together, these various modes (capitalist production and exchange, welfare economy, including education, healthcare and welfare arrangements, household economy, voluntary work) form the economic dimension of reproduction of present-day capitalist welfare states. Such an anthropological conception of the economy is inspired by Polanyi.[2] This economic dimension is of prime importance for reproduction, and it co-determines the degree to which polities grow or decline. In addition to the modes mentioned above, it also presupposes general conditions such as an environment that enables the activities. However, it should be evident that this economic dimension is not really sufficient. Indeed, we have indicated the various social identities that go hand in hand with the historically variable forms of organizing the interrelated economic facets of present-day polities, and these identities point to other dimensions — social, political, cultural and ethical ones — that are also necessary for reproduction.

How these various dimensions are related and connected to each other, and what their impact is in the reproduction of polities is historically variable. Moreover, the reproduction can follow quite different paths in the various polities that comprise the world system. It should, however, be clear now, that when we refer to the dynamic process of reproduction, we consider the mechanisms of reproduction not as machinery that produces identical results independent of the starting positions or the materials used. Indeed, if the mechanisms of reproduction are (re-)constituted by facets of the existing dimensions, such as economic or cultural ones, there is, of necessity, a general form of path dependency that is much more universal than the purely financial or institutional path dependency that is used in policy studies.[3] This formulation is inspired by the work of Taylor on forms of modernization.[4] Taylor distinguishes cultural and a-cultural theories of modernization. The dominant theories are the a-cultural ones, pretending that modernization, characterized for example by industrialization, by city-forming, by rationalization, etc., produces the same outcome independent of the starting positions. Taylor rejects such a view, showing that existing worldviews, including conceptions of persons, of societies, of the division into economic, moral and other dimensions, together with the actors involved, have a significant influence on the process of modernization. Therefore, he concludes that we are confronted with a multiplicity of processes of modernization, and that one can, for example, speak of western, Chinese, Indian and other processes of modernization. This passage can therefore be read as a generalization of Taylor's work.

We can summarize the preceding remarks in the following table, which presents in a simplified way the various economic modes and facets of capitalist economies, their variations and transformations. This table circumscribes exclusively the economic dimension and not the other dimensions, such as social, political and ethical ones.

We restrict ourselves to polities that are capitalist welfare states, organized on the basis of the rule of law and with a (representative) democratic

Table 1. Economic modes

Economic mode	Facets	Variations	Historical transformations
Capitalist production and exchange	Specific production processes, various markets (labour market, financial market, etc.)	Variations depending on: environment, sector, qualifications, etc.	Overall growth, including important crises
Welfare economy, including education, healthcare, welfare arrangements	Specific welfare arrangements such as pensions, formal care systems, education system, etc.	Considerable variations between polities, depending on political actors, requirements of the production system, etc.	Overall growth during last century, with important differences between specific polities
Household economy	Informal care (for children, the elderly, sick people), daily care (cooking, cleaning, etc.)	Depending on the composition of the household, degree of mechanization of household tasks, existence of productive activities, distribution of tasks (man/woman)	Significant rationalization and simplification in European polities during last century, changing composition of households
Voluntary activities	Informal care, neighbourhood activities, etc.	Numerous	Changing intensity and focus

form. As such, the different aspects of citizenship (usually the so-called civil, political and social rights, see Ref. 5) come into play. There is certainly a historical movement of generalizing these aspects of citizenship in other parts of the world, which has been rather successful, without, however, the result that citizenship rights have been introduced in all polities in the same way, nor to the same degree. A simple generalization would indeed contradict the path dependency we have formulated.

We now have all the necessary elements for formulating the central question of this contribution. How is it possible to realize, particularly in European capitalist welfare states, legitimate forms of welfare arrangements, and pension systems in particular, within the ongoing dynamic transformation of the world system? One should not expect that legitimate forms of such arrangements, situated mainly in the welfare economy, are completely independent of the other economic modes, in particular of the household economy and the mode of capitalist production and exchange. The criteria for specifying 'legitimate arrangements' are formulated taking as the basis the citizenship norms that emerge from the dynamic of the development of citizenship. Indeed, as outlined, the changing forms of citizenship are connected to other dimensions. In European capitalist societies in particular, citizenship is linked to the economic dimension as has been argued above, for example through the presuppositions that: (1) citizens are self-responsible for their lives (in the past strongly related to the principle of subsidiarity); and (2) individuals are endowed with several rights and duties, such as the right to autonomously conclude contracts. Therefore, there is a natural link between norms derived from citizenship, and the economic dimension of which welfare arrangements constitute one specific mode. In this contribution we limit ourselves for the purpose of analytical clarity to these two dimensions, and refer to other dimensions (such as political and social ones) only briefly where necessary.

We first formulate the norms to be used (Sec. 2). Secondly (Sec. 3), we demarcate the arena of exploration, such as the changing types of households, in which the various social identities of employee, consumer, carer, etc. have, in recent decades, undergone profound transformations. Moreover, we address the question of how European pension systems were related to a specific historical constellation of household economies, and how the recent pension reforms attempt to adjust the pensions systems to new configurations of household economics. In a third step (Sec. 4), we then indicate why the reformed pension systems do not fit well with the changing constellation of households, using the norms formulated. We then (Sec. 5) address the question of an 'ideal' pension system that (Sec. 6) would be in line with the formulated norms. Some final remarks (Sec. 7) serve to qualify our 'realistic utopia'.

2. Norms Derived from the Dynamic of Citizenship

Citizenship has factual and normative dimensions.[a] Factual components include, for instance, the established citizenship rights that are effectively applied in the present circumstances. This means that well-known procedures and institutions do exist, which permit the enforcement of rights should they be threatened in any way. One can take as an example the right of women not to be discriminated against. For a long time this right was a possible ideal norm, without it, however, being put into practice. At present, however, there is a growing number of European polities that do enforce this right, but with significant variations concerning the social space where the right is enforced, such as the workplace, the family, public space or one's own body. This example also enables us to illustrate what is

---

[a]Marshall also speaks of an 'image of an ideal citizenship against which achievements can be measured and towards which aspirations can be directed' (Ref. 5, p. 29).

meant by the normative dimension. Indeed, from the very beginning, non-discrimination of women was for some a normative ideal of citizenship, even if this ideal had been contested by many in the past (and even sometimes in the present).[6]

Citizenship can be seen as something that is static and dynamic at the same time. Considered from a static point of view, citizenship can be defined as a set or register of established rights and duties, including procedures, for example legal ones, that guarantee that these rights and duties are adhered to. Such an idea is similar to what has been termed the factual dimension of citizenship, but which is now seen from a different point of view. Following Marshall,[5] one can distinguish civil, political and social rights and duties, but various authors have indicated that this list is far from complete. Held,[7] for example, argues that 'political-economic rights' must complement this list, and Isin and Wood introduce many other rights,[8] such as reproductive rights, ecological rights, rights of individuals with a functional limitation, rights of homosexuals, etc.

However, there is also a dynamic conception of citizenship, apprehending the historical movement of struggle of establishing or expanding rights.[b] Hirschman,[9] for instance, has studied some episodes of this struggle. He studied the arguments used against the introduction of civil, political and social rights, and he succeeded in showing that these arguments do not reject the rights as normative ideals, but try to reject or limit the introduction of new rights, arguing that this introduction would, given 'human nature', not lead to essential change but only threaten the rights that had already been introduced. In the last two centuries, for example, the argument that the generalization of political rights would threaten the established civil rights (the influence of the uneducated majority will undermine the liberties obtained thanks to the established civil rights) had some considerable influence. In short, the dynamic of citizenship points to the historical struggles extending, and sometimes reducing, the range of citizenship rights and therefore the areas where new, or reduced, forms of participation of individuals as citizens are possible in organized societies.

The static and the dynamic version of citizenship both include normative perspectives, the first in terms of ideals of realization of rights, and the second as ideal forms of participation to be achieved in the ongoing struggles extending existing rights or establishing new rights.[9] The first meaning points to the degree of realization of a given right. The example

---

[b]This corresponds to the Weberian understanding of social policy as cultural struggle.

used here — the right of women not to be discriminated against — was, as said, only gradually introduced with considerable variations of the social spaces involved: The right to vote and to be elected was introduced during the last century, but once introduced, this right was enforced to a high degree. In other areas, such as the workplace, the introduction of this right was only achieved much later, and applied less systematically than in the political space. The second meaning points to normative ideals of full participation in the polity, which can be constrained even when specific rights do exist but are not applied in a pertinent way. As we point out in later sections of this contribution, even if women have obtained the right of equal participation in social arrangements of the welfare state,[c] the problematic application of this right in the case of pension systems severely constrains the opportunity for women to obtain pension entitlements similar to those of men.

If we view citizenship from a dynamic perspective, then we focus on the struggles concerning the establishment of new forms of practices and opportunities for members of the polity. It is not possible to isolate this dynamic from the overall dynamic reproduction of polities, and in particular, from the struggles in the economic dimension. Indeed, new practices and opportunities for action have to be enabled, also materially. These practices and opportunities either presuppose time, space or other resources that have to be provided by one or other economic mode. Quite often, the welfare economic mode will be the first candidate to provide the resources, followed by household economies, voluntary activities and capitalist forms of production and exchange. This means that the shaping of citizenship is not, and cannot be, a linear process, because in times of economic crisis (of the classical variety) the interdependence of the various economic modes will entail a reduction of the resources provided, and therefore also a possible retrenchment of the corresponding citizenship dimension. Take, for example, the fluctuating development of health, education or social provision over time and in different countries. The general tenet of this statement is that there can be advances but also retrenchments in the citizenship arenas.

Such an approach means that citizenship cannot be seen separately from the other changing dimensions of the historical dynamic of polities. In particular, this approach cannot refer to norms as starting points of

---

[c]Throughout history women have also benefited from positive discrimination, particularly in pension entitlements. Accompanied with the emphasis on a unisex life-course, these positive measures for women are on their way to being reduced.

the analysis.[10] On the contrary, norms are conceived of as elaborations of the various groups that comprise the polities, including their associations, going beyond the boundaries of any given polity. And this implies that at any moment there are multiplicities of norms at stake, elaborated by the different groups in different polities.

We limit ourselves in this contribution to European polities, and to the norms developed in terms of a static and dynamic version of citizenship.[11] These norms can be and are formulated in terms of ideals of participation of citizens in European polities. However, we will not enter here into the traditional discussion of how one can justify citizenship. There are elaborated justifications from a liberal point of view, from a communitarian one, and there is also a republican version of the justification of citizenship. We will certainly use much of the liberal elaboration, but we recognize the necessity for the limited (to individuals) rights of a liberal conception to be generalized and guaranteed for all men and women living in these polities, such as for example the right of equal opportunities. Moreover, from a republican conception of citizenship, other ideal types of participation are important, such as the links of the individual members to the collective, not only within the various European polities but also Europe-wide. In particular, the so-called democratic deficits within the European polities and their association in the EU have to be overcome when implementing such ideals of participation. This also entails taking account of the situation of migrants and EU interrelations with other polities in the world system.

The traditional citizenship norms — liberty, equality and solidarity — are quite well-known and reasonably precisely formulated. However, certain national and economic interests currently hinder and constrain the realization of these norms in European polities. Moreover, these normative ideals are subject to different interpretations by the various political forces in Europe. Finally, the inter-connections between these rights can be seen in various different ways. For example, for some actors freedom is the primary normative ideal, whereas others give preference to equality. And solidarity is, however, a somewhat contested dimension of citizenship as we will see when exploring pension rights.

Recognizing these debates and confrontations does not mean that a more systematic elaboration of these norms in the domain of welfare arrangements, and in a particular one, namely pension policies, is a futile exercise. On the contrary, such a conjecture allows one to point out the concrete lines of debate and confrontation and the various interests involved.

For the sake of simplicity, we take the traditional norms as our point of departure, and in the following table illustrate the controversies that arise when trying to apply and develop these traditional norms for actual pension entitlements.

Table 2. Citizenship norms and actual pension entitlements: A first illustration

	Freedom	**Equality**	**Solidarity**
Specification	Criteria and limitations to build up pension entitlements	Conditions of building up pension entitlements	Different forms of inter- and intra-generational solidarity
Contested points	Waiting periods, Mandatory retirement age, Labour market relatedness/dominance, Statistical discrimination (LM), Institutionalized role differentiation	Reasons for establishing entitlements, such as childbirth, care, learning, etc. Life tables, Retirement age	Abusive forms of solidarity (e.g. from the poor to the rich), Inter- and intra-generational solidarity, and Inclusion/exclusion (e.g. migration)
Economic modes	All the various economic modes, excluding the voluntary activities, are, varying per country, important for enabling the building up of entitlements	Involves the various economic modes, and entails a critique of the established boundaries between the various modes	Primarily the welfare economy, providing guarantees concerning various transfers (based on contracts and legislation)

3. Preconditions and Characteristics of Pension Systems

Pension systems implicitly instituted life-course norms. Therefore, the development of pension systems also needs to be approached through life-course. As stated above, modern capitalist economies are multi-modal, and depend on non-market and non-commodity reproduction as well as markets and production. Production, exchange, distribution, consumption and so forth are traditionally allocated to different phases of the life-course divided up into: (1) socialization and learning, (2) labour market participation, and (3) retirement. These 'ages'[12,13] are subject to fundamental transitions[14] whilst the development of the 'third age', i.e., retirement, is based on the first two ages and depends on a historical and specific link between the different ages.

The production and circulation of the various resources, including the resources for financing pension entitlements (in fully-funded systems, FF, and pay-as-you-go systems, PAYG, alike) is an inherent component of capitalist logic, however differently instituted. Therefore, the dynamics underlying the flows of resources are of great importance, and cannot be captured by the typologies of welfare regimes, such as the well-known regimes elaborated on by Esping-Andersen.[15]

Pensions form a very peculiar resource due to the fact that they cannot be conceived of within a conception of classical property rights. Although PAYG and FF systems are quite different, in relation to timescales as well as property characteristics, all pensions depend on economic growth. This kind of dependency, which is more than individual short-term economic survival, is typical of modern capitalist systems.[16] It is linked to changing, and improved, conditions of the quality of employment relations. In addition, it depends on a specific kind of trust in the stability of arrangements, to be organized for instance in insurances representing a broad social basis.[17]

Equilibrium is necessary to be able to finance the first and the third age, i.e., proportionality between the different forms of social reproduction of the various modalities of economy. It is self-evident that, for various reasons, the dynamic character of the economic sub-systems can bring about changes of the 'typical ages' of the life-course. The linkages and interdependencies between the various phases of the life-course are redefined, too. Therefore, the characteristics of pensions as the third age are also on their way to being transformed.

The characteristics of pension systems are generally rather similar, although their concrete institutionalization varies widely per country. Pension norms are, in one way or another, directly linked to age and labour market participation. This is not surprising since pensions, in all countries, are a means of regulating the labour force. And *vice versa*, the dynamics of labour markets also find expression in reforms of pension arrangements.

Understanding the consequences of pension reforms is quite challenging since pensions are not only a very complex nexus of institutions but are also spread out over different timescales that go beyond short-time planning. Pensions, therefore, are a kind of intergenerational contract: while entitlements have to be built up in the 'second age', using this terminology, pension benefits, in terms of monetary means or services, are enjoyed in the 'third age'. This means that rights over resources have to be conserved over time.

Entitlements need to be built up. The number and the weight of reasons for building up pension rights vary country by country. Based on these differences, the pension systems in different countries establish unique identities. These identities depend on instituted normative life-courses and their further development in practical terms. All modern capitalist systems, however, have implemented pension regulations that allow for a period of non-employment, which is in line with the capitalist logic of spreading resources over time. As a consequence, pensioners are, at least in theory, enabled to enjoy a period of non-employment with a specific level of independence. They are, in financial, and ideal terms, not directly dependent on either their children or on kinship solidarity. They merely depend on societies performing their instituted obligation concerning the rights built up in the past.

4. Male Breadwinner Systems and (Limited) Individualization: Non-Fit of Pension Systems

The life-course norms implicitly instituted in pension systems established during the last century are currently being transformed. In most European countries, pension entitlements were calculated following a highly gendered 'life-course regime',[18] representing the perspective of a household comprising a female housekeeper and carer and a male breadwinner, the so-called male breadwinner model.[19] While women in traditional heterosexual legalized household formations mainly stayed at home, i.e., in the 'private' domain, in order to take care of the 'reproductive' necessities of this unit (housekeeping and caring for the elderly and the children), men earned the 'family wage' within the 'public' domain in order to provide for the family.[20] As a logical consequence, social rights were also related to the 'family wage' and this 'male breadwinner' social model.

Not only were social rights related to the family. Civil and political rights, to use Marshall's categorization, were also instituted on the basis of the family unit and the male as head of this unit. And although political rights, as for instance the right to vote, and civil rights, as for instance the right of women to sign contracts, changed, other civil rights have only recently been generalized. The so-called private domain, for instance, qualifies after partly putting problematic issues such as the level of education or domestic violence in the public domain. And social rights in particular are still in the process of transition due to their behavioural component and the process of individualization.

Various developments have led to a more individualized understanding: the traditional family formation lost its self-evident status; women participate in the labour market more extensively; discriminatory legislation began to be reduced (for instance, it was forbidden in the Netherlands to employ married women in most sectors until the 1960s). However, this strong impulse of the 1970s in particular did not adequately find its expression in, for instance, institutional change. Although it is correct that much legislation changed in many countries, institutional legacy and neo-traditional trends hampered the full individualization of both welfare arrangements and the possibility of complying with the unisex life-course norm. It is debatable whether full individualization would be desirable or even possible. However, it is obvious that the transitions of welfare arrangements that have been implemented eliminate the gender distinction insufficiently. The new, recently established life-course norms, which require individual ideal labour market participation, do not yet fit well with the life-courses of women. In combination with other welfare transitions, the disadvantages of this non-fit exacerbate the situation. This is the case since social entitlements, mainly, if not exclusively, built up within the domain of the market economy, are under pressure, and welfare arrangements are subject to retrenchments and individualization.

While men may also have difficulty complying with the normative life-course, particularly in times of the 'new social risks' of rising unemployment and transitional labour markets, women face many more obstacles than men do.[21] This is the case for various reasons, including:

(1) labour market segregation — across the board and also in old age and part-time employment, in wages, in functions, in life-long learning;
(2) care responsibilities — and the institutionalization of parental leave, care facilities, care credits;
(3) tax systems — advantageous for 1.5 earner households (and pensions) in most European countries;[22]
(4) statistical discrimination — based on possible care-related time off and life expectancy.[23]

What kind of individualization could be envisaged in order to work out a more legitimate but yet practicable pension and welfare organization?

5. Realistic Utopia

After analysing the norms, characteristics and shortcomings of pension systems, we now outline a utopian pension system based on realistic developments. We refer to it as utopian due to the fact that it is highly unlikely that a system like this could ever be realized — at least in this particular way and at least in the foreseeable future. Such a system is utopian for two reasons: (1) it would need ideal circumstances, for instance in the labour market and in information flows, and (2) the dynamic nature of systems subverts the conservation of a status quo and continuously requires adjustment, adaptation and modulation. In addition, although our pension model is much more equitable than all the existing pension systems, it is still based on problematic conditions, as we point out when describing the different layers. We say it is realistic because we base our scenario on actual, partially achieved, solutions to the pension question introduced in different European countries in the past few decades. Some of those are understood as norms, others as concrete measures, and all of them are formulated as being necessary for reproduction by different actors. Such a thought experiment follows the tradition of humanist thought experiments of Thomas More or Erasmus van Rotterdam; among recent publications it can be compared with the 'postindustrial thought experiment' of Nancy Fraser[24] or the 'new model' of Stephan Lessenich and Matthias Möhring-Hesse.[25]

Reasonably legitimate pension systems, we argue, can only be based on the acknowledgment of the preconditions of societal sustainability and capitalist productivity,[26] or of what we called reproduction. Pension systems, therefore, should be based on three 'layers', i.e., pension entitlements should be linked to the following three conditions and requirements:

(1) poverty prevention;
(2) valuation of necessary contributions to society (social reproduction and care, life-long learning), and
(3) valuation of labour market participation (economic productivity).[d]

Each system should, in principle, include at least all these three layers in their pension system to reflect a balanced relationship of different tasks and responsibilities. We call them layers for two reasons: firstly, to avoid any association of 'our system' with the differentiation of pension schemes or pension pillars, and secondly, to explicitly characterize the pension reasons as complementary: on the first layer of universal old age pensions, the second layer (credits for care etc.) and/or the third layer (labour market related contributions) can and should be additionally built up.

The complementary structure is inspired by the Dutch 'cappuccino mix' with its basic pension for all (the coffee), its semi-compulsory and complementary occupational pensions for most (the milk), and the possible

---

[d]Other valuations may be included. Military service, for instance, may form part of layer two or three, depending on the country's (military) system. Private savings are more a form of ownership than part of pension systems. Some countries are inventively including them in overall pension resources through taxes and regulations.

Table 3. Complementary pension layers of a realistic utopian pension system

	Eligibility	Function	Source of resources
1st layer	Universal, not means tested (e.g. residency based)	Poverty prevention (e.g. social assistance)	Public purse, General revenue
2nd layer	Social contributions, Human capital, Solidaristic elements	Complementary additional pensions (credits)	Public purse, General revenue
3rd layer	Financial/economic contributions, Real capital, Proportional wage-related calculation	Complementary standard securing pensions	Proportional wage-related contributions[e]

complementary private additional pensions for those who can afford it (the cocoa), despite all the important differences.[27]

These layers would, for democratic and sustainability reasons, need to be based on the general principle of justice in inter- and intra-generational terms. This means that the flows of resources are discussed, related to the various kinds of activities, related to the different economic modes.

5.1. *Basic Pensions*

The first pension layer (1) we describe is inspired by the pension systems of the Netherlands and Denmark. These systems comprise a universal basic state pension as the cornerstone on which further pensions may be built up.

Poverty prevention is one of the basic intentions of welfare states, enabling the participation of individuals. Although poverty levels vary considerably country by country, all European welfare states do have a social assistance institution to guarantee a minimum of individual autonomy and political and social participation. It is not the aim of this article to contribute to the discussion of general basic income; we focus on old age only. Due to the fact that labour market policy restricts labour market participation to a specific age (the variable mandatory retirement age), it is inherent in the system that the resources necessary for living after that age be delivered. It is in the interests of society as such that the elderly residents, who are, institutionally or physically, unable to earn their own living, also have

---

[e]Labour market activities contribute to productivity which is necessary to maintain wealth also in the wider sense. Therefore, it could be argued that these activities should similarly be financed by the public purse. However, to disconnect the linkage between wage and status securing social rights might form part of a utopia, not, though, of a realistic one.

the guarantee to be full (political, civil, and social) citizens. This minimum needed to maintain a decent standard of living should therefore be financed by general revenue. Furthermore, pension resources are one of the cornerstones of resource flows.[f] In addition, such a guaranteed universal basic pension would benefit the generation contract that forms a basic condition for capitalist welfare states.[29]

As a matter of fact, a universal basic pension would, at the same time, solve another urgent problem of recent pension systems: that of trust.[30] This trust can be seen from two perspectives: firstly, all employees would feel a kind of self-evident security in old age and therefore they would more readily invest in additional resources for their old age (see the Netherlands and Denmark; the reluctant acceptance of the German *Riesterrente* may serve as an example of low levels of investment due to little trust in pension investments). Basic pensions that are meant to just prevent actual poverty, are unlikely to discourage additional savings. Secondly, in times of unstable labour markets, such a guarantee for old age security for all contributes towards social cohesion and, thereby, it contributes to trust in the economic and political systems;[31] it would not, however, fundamentally reduce labour costs.[g]

Surely the level of benefits of such a basic pension is subject to political considerations, and will therefore always be in danger of being reduced. However, if societies take this democratic institution seriously, basic pensions could get the status of a self-evident element of democratic systems. Yet one should not underestimate the dynamic of welfare arrangements: economic growth can lead to stronger developments of income-related additional pensions and by that, put the legitimacy of basic pensions under pressure (the developments in the Netherlands resulting in retrenchments of basic pensions may serve as an example). In Denmark, due to a strong pensioners lobby, the basic pension is, although reorganized, under hardly any pressure. It should be stressed that pensions, in particular, are positioned in the field of tension between influential political powers, that have recently gained influence, holding a neo-liberal perspective on the one hand, and the specific long-term characteristics of pension systems as institutions

---

[f]In 2001, the share of pensions in GDP was 12.5% in the EU25.[28]

[g]As already pointed out by Sinn,[29] shifting financial burdens does not reduce them. Financial means are and will be transferred through the fiscal constitution charging productivity and growth still mainly through wages. However, universal pensions would reduce the enormous costs related to survivor's benefits and widow(er)'s pensions. Furthermore, universal pensions would involve relatively low administration costs.

on the other. Although opposition to social reforms has generally increased in the past decade, political resistance to pension reforms is less strong than resistance to other social reforms due to the fact that they will have the most important consequences for future pensioners who might not yet feel concerned.

In the Netherlands and Denmark, this universal not means-tested basic pension layer is related to residency. It could be argued that, by buying services and paying different kinds of taxes, such as VAT, residents do contribute to the public budget and to the circulatory logic of welfare economies. It remains, however, problematic, in how far transnational labour transitions or migration in general hampers the opportunity to finally build up more than a basic pension (the so-called pension gap, see Ref. 27). Sophisticated solutions have to be found in times of increasing labour and capital mobility.[h]

5.2. *Complementary Pensions*

Other layers, with further conditions for entitlements based on contributions to sustain the polity, should complement the first one, which is meant to be the basis for social cohesion. The contributions to the polity may be financial, economic and social. Sinn for example, uses different terminology and speaks of investments based on 'real capital' and 'human capital'.[29] Institutionalizing ideal concepts, such as the Dutch combination scenario, would result in each citizen building up pension rights in both complementary layers. However, this depends on the definitions of what is seen as (social) contribution and on the concrete life-course of citizens.[14] The relationship between second and third layer pensions would, therefore, depend on various facets of the systems and individual decisions. It is unlikely that future pensioners will not have participated at all in the paid labour market, i.e., only built up care credits. Nor is it likely that each future pensioner will be able to build up full occupational pensions in the current understanding in times of very flexible and insecure employment records. The traditional recipients of labour market related pensions and the traditional understanding of its level are on their way to being transformed. Therefore, the concept of reorganized pension reasons also reflects developments within traditional pensions which provide sufficient practical grounds to base all citizens' future pensions on complementary pension reasons.

---

[h]Categorizing resources is difficult. Ferrera sees territory as the means that made it possible,[32] whereas Durkheim underlines interdependency.[33]

5.2.1. *Second layer pensions*

One of these complementary layers, the second layer of pension entitlements (2), would give value to societal contributions in the sense of 'solidaristic elements' (Ref. 34, p. 167). This second layer is inspired by some countries' institution of pension credits for care or for education (France, Germany, Austria). In addition, it reflects some researchers' ideas on even the pure economic legitimacy of such socialization of financial burdens, arguing that childcare in the end pays for itself[35,i] or that such human capital investment has to be seen as part of the generation contract inherent in all pension systems that needs to be financially valuated in addition to direct financial investments.[29] In this sense, we formulate this part of a consistent logic of individualized pension systems. Highly individualized systems tend to face the Prisoners' Dilemma, i.e., everybody tries to avoid personal disadvantage (specifically, a loss of wages as a result of parental leave, for instance) while being aware that somebody else has to take responsibility for actions that entail disadvantages.[38] The second layer pensions therefore try to overcome the seeming dichotomy of self-interest versus solidarity.

If the generation contract is a precondition for sustainability, new generations are a condition for reproduction and these generations need to be educated in the knowledge societies of our historical and geographical situation. While formerly 'familialized' systems are being defamilialized in terms of social rights, pension reforms do not adequately transform former family rights into new ones. When 'reproduction of society' is recognized as a 'social purpose' (Ref. 39, p. 250), related costs may legitimately be financed by means of general revenue.[29] In addition, pension credits for 'socially useful' contributions (Ref. 40, p. 105) should be implemented as non-conditional entitlements, i.e., they should not be linked to means tests or entitlements gained by labour market participation. As an addition to basic pensions, they value this societal contribution. This unconditional valuation of child care credits was the case in France until 2004 (recent reforms introduced conditions related to labour market participation).[j]

---

[i]Concepts of equal rights need to be accompanied by concepts of equal obligations to children and other dependents.[36] This problem is also exposed in the concept of 'inclusive citizenship'.[37]

[j]We disagree with Ferrera and Rhodes who call for "incorporating 'equitable' and 'updated' norms in the crediting of contributions for involuntary or socially valued interruptions of work (for example, training or caring periods)" (Ref. 31, p. 269), since this conditionality of credits on work interruption firstly contradicts the *per se* valuation of socially useful contributions and secondly hampers women's equal positioning on the labour and skill market.[41]

As stated above, economy has to be understood as an interplay of its four interrelated modes, including the social reproduction of labour, encompassing education and care. It is challenging to specify a list of socially useful contributions and to institutionalize a connection with specific pension entitlements. It is not the intention of this contribution to deliver such a list. Yet we can point to some possible components by analyzing academic and political attempts to specify them.

Some countries have introduced care credits into their pension systems. Although it is correct that only those countries without universal basic pensions implemented such credits, there is no convincing argument to justify this exclusive linkage. On the contrary, a holistic approach adapted to the anthropological concept of the economic dimension values necessary contributions to the system apart from those related to the market economic mode. Actual pension care credits, in different countries, entitle one to different rights in terms of: (1) absolute (fixed) or relative (income related) monetary means; (2) full-time, part-time or non interruption of labour market participation; (3) the duration of building up care-related entitlements; (4) the number of children cared for, and so forth. In some countries, there are additional care credits for parental care of handicapped children.

Another socially useful contribution discussed (and partly introduced in the UK or as a possibility in Germany and the Netherlands) is geriatric care. And there are of course various additional useful time investments which could be thought of. This discussion tends to lead to the approach of the so-called third sector.[40] However, the similarity may concern the valuation of activities, not the categorization of them. Furthermore, education credits are (very differently institutionalized) part of some countries' pension systems.

To conclude, what we call the second layer would serve social cohesion in two ways. Firstly, by contributing to the gender contract (not in the sense of gender roles but in the sense of a more or less 'just' valuation of different activities, and not exclusively the activities in the market mode), and secondly, by contributing to the generation contract. Therefore, the costs of this layer should be covered by general revenue and not by specified groups only (through e.g. payroll taxes charging wages only).[k]

---

[k]The realistic utopian pension system needs, as existing pension systems do, further elaboration of (new) financial responsibilities and a critical analysis of resource flows in general. For instance, which sources can be tapped for the national budgets as general revenue in times of supranational economies?[14]

5.2.2. *Third layer pensions*

The third layer (3) of our pension utopia already exists in all European countries: labour market related pension entitlements. Since we analyze the meaning of different pension reasons, we do not differentiate between occupational and private pensions or between PAYG or funded pensions, presupposing that firstly, both resources are related to the labour market as a financial resource or as a commodity in the widest sense, and that secondly, the management of these financial contributions is very controversial and of rather secondary interest when comprehending this layer as complementary. As resource flows are mainly related to the market economy and as productivity in the wider sense is necessary to maintain wealth, labour market contributions need to be valued as societal ones do. For most researchers, this is rather self-evident, so we will not discuss this kind of legitimacy any further.

What we do want to discuss in more detail is the problematic preconditions and calculation norms that determine how such entitlements are built up. As Ferrera and Rhodes put it:

> "The elimination of transfers that can be identified as inequitable (because they are grossly disproportional to contributions), outdated (because they are out of step with the structure and distribution of needs) or perverse (because they generate significant work disincentives) appears desirable both for normative and practical reasons." (Ref. 31, p. 268)

We will focus on the following issues: (1) proportionality of contributions and entitlements; (2) identities of the insured; and (3) security of the investments.

(1) Calculation norms may lead to disproportional benefits in relation to contributions. Differences in wages and in the quality of contracts are rather 'normal' in capitalist welfare states. However, there is no legitimate reason why calculation norms of pension entitlements should even extend these differences. Nonetheless, this is the case in many pension systems. In the Netherlands, for instance, double wage income may result in four-fold pension entitlements.[42] A comparable effect can be observed in France where the calculation formula, based on time, may lead to a 50 per cent loss of pension benefits when one pays 12 per cent less contribution.[43] This is an extreme example of a malus regulation; and a lot of bonus–malus regulations have recently been introduced in different countries.[44] In addition,

several countries' pension systems include specific tax regulations that are generally more beneficial for higher income earners. Such additional disproportionalities lack legitimacy.

(2) Moreover, different recent pension reforms partly undermine the nature of pensions as an insurance by transforming them into individual account systems. If, however, pension arrangements change their meaning from old age insurance to (individual) old age savings, social cohesion is under threat.[17] This is even more the case in times of unstable labour markets. All contributions to economic productivity should be included in this 'third layer insurance': those by the self employed, by low wage earners, lawyers and labourers — just to name a few — alike. This is not yet the case due to a variety of conditions (qualifying periods, groups specific pension schemes, and so forth). Some countries are changing these conditions as well as unifying their pension sub-systems, and, concerning membership, extending their differently separated systems (e.g. Germany, France, Denmark).

(3) More individualized pension investments entail more individualized insecurity. Scandals in Britain and the US, for example, have served to indicate the powerlessness of people when faced with big firms going bankrupt, which completely annihilated the companies' huge pension responsibilities.[12] An opposite example is the highly regulated and collective pension funds in the Netherlands. Obligations of pension funds are strictly regulated so that comparable disasters are unlikely to happen. The Dutch system also manifests considerable fantasy by combining calculation methods, such as 'defined benefit' systems and 'defined contribution' systems. Such pension formulae are in line with some academics' perspectives. Myles for instance, argues against fixed rates in general, in favour of flexible rates necessary to sustain PAYG systems, which are more suitable when it comes to creating trust.[34] In our opinion, pension funds, or indexation in general, could even be regulated in a more sophisticated way, combining overall and long-term profits (and losses) with a more general redistribution.

To conclude, the three issues of 'third layer pensions' (proportionality, identities of the insured, and security of investments) can be handled more democratically if legally instituted, i.e., regulated, in the pension system as such.

6. Coming Full Circle: Citizenship and Pensions

Pensions as a kind of 'societal property right' institute a socialized ownership.[45,46] Therefore, pensions should be in accordance with the norms of citizenship. Regarding pensions from such a perspective, we have come full circle: pension rights as part of social rights belong to citizenship rights, although continuously contested by various power struggles. However, compared with its political brother and civil sister, pension rights as specific social rights, are barely understood and have not been discussed in this light up until now.

Our realistic utopia does not come close to any pension system classification as, for instance, expounded by Bonoli (Bismarckian or Beveridgean systems) or by Pierson (categorizing pension systems into those that moved PAYG to FF before the system matured and those that didn't), or others, also due to the fact that our layers do not correspond to pension schemes or pillars. Our attempt has been to formulate a holistic concept of welfare systems, recombining democratic citizenship (rights) and long-term necessities of economic and polity sustainability.

Table 4. Citizenship norms and realistic utopian pension entitlements: An illustration

	Freedom	**Equality**	**Solidarity**
Specification	Trust, social cohesion, security of investments	Valuation of activities in different economic modes, proportional benefits, inter- and intra-generational justice	inter- and intra-generational solidarity
Contested points	Role of the state, regulations	List of socially useful activities, concrete value of basic pensions and of different pension credits, indexation (redistribution)	Transitional labour markets, financial responsibility of trans-national companies, migration
Economic modes	Basic freedom via welfare economy, additional investments based on financial means (market economy)	Credits for socially useful activities via welfare economy, additional investment equally valued via rule of law, all modes included	Welfare economy and regulations, more fantasy to establish new sources and to combine resources

This realistic utopian pension system combines what is fragmented or partly even ignored in present-day pension systems: the different economic modes and the traditional values of citizenship, including the long-term sustainability of capitalist welfare states. The dichotomy between the public and the private is the crux of welfare systems in gender terms. Another dichotomy is artificial, namely that of (individual) self-interest, including freedom, and solidarity, sustaining social cohesion. The art is to find equality in opportunities to comply with different life-course tasks to solve the Prisoners' Dilemma of our systems. The art is also to provide choices to combine elements of different life-courses without the risk of living in poverty in old age.

Citizenship rights, which are defined as individual rights, have to be combined with citizenship duties in the sense that the reproduction of the polity is not solely related to economic growth but to all the four modes of the economy. The different essential activities related to these economic modes, summarized by us elsewhere in the term 'activeness', are therefore different from the concept of 'economic citizenship' which sticks to the problems correlated with the concept of commodification.[47] Our understanding of citizenship and the social identities involved, entitling one to various rights, is a holistic, anthropological, long-term sustainable one.

7. Final Remarks

We are aware of the fact that this text has many loose ends. We do not pretend to weave them together in these final remarks. We limit ourselves here to indicating a number of important points for further reflection.

Firstly, the 'realistic utopia' we formulated here for pension systems cannot be used without many changes being made for other welfare arrangements.

Secondly, we already underlined the interdependence between welfare arrangements and other economic modes, and more generally, with other dimensions, such as political ones. For example, we did not address at all the process of political decision making.[1] Neither did we work out in any detail the various interdependencies between pension systems and other welfare arrangements, the relationships between pension systems as belonging mainly to the economic mode of the welfare economy and other economic modes (with the exception of the household economy), nor did we address the difficult question of how our conception of norms (derived from a dy-

---

[1]Debates on decision making and implementation of different norms are numerous. One might think of the exchange of Nancy Fraser, arguing on her concept of 'participatory parity', and Axel Honneth, comprehending social policy as little steps towards utopia.[48]

namic version of citizenship) is related with the worldview we used when introducing the question of reproduction of present-day capitalist societies.

Finally, we have limited ourselves to designing a realistic utopia concerning pensions for European polities. But at the same time, we recognized the existence of a significant interdependence between all the polities in the world system. Therefore, we cannot pretend that the realistic utopia suggested here has any autonomy. On the contrary, we are fully aware that only a theory of the worldwide dynamic of more or less uneven reproduction and transformation of the world system would permit us to indicate the limitations of our work and to go beyond these limitations.

References

1. I.M. Wallerstein, *World-systems Analysis: An Introduction*. Durham, N.C.: Duke University Press (2004).
2. K. Polanyi, *The Great Transformation. The Political and Economic Origins of Our Time*. Boston: Beacon Press (2001). [originally published in 1944]
3. J. Myles and P. Pierson, The comparative political economy of pension reforms. In: P. Pierson (Ed.), *The New Politics of the Welfare State*. Oxford: Oxford University Press, 305–333 (2001).
4. C. Taylor, Two theories of modernity. *Public Culture* **11** (1), 153–174 (1999).
5. T.H. Marshall, *Class, Citizenship and Social Development*. Westport: Greenwood Press (1964).
6. R. Lister, *Citizenship. Feminist Perspective*. New York: University Press (2003).
7. D. Held, Citizenship and autonomy. In: D. Held and J.B. Thompson (Eds.), *Social theory of modern societies: Anthony Giddens and his critics*. Cambridge: Cambridge University Press (1989).
8. E. Isin and P. Wood, *Citizenship and identity*. London: Sage (1999).
9. A. Hirschman, *The Rhetoric of Reaction: Perversity, Futility, Jeopardy*. Harvard: Harvard University Press (1991).
10. K. Olsen, *Reflexive Democracy. Political Equality and the Welfare State*. Cambridge/London: MIT Press (2006).
11. R. Maier, Quality of citizenship and interdisciplinarity. In: R. Maier (Ed.), *Citizenship and identity*. Maastricht: Shaker Publishing, 181–208 (2004).
12. R. Blackburn, *Banking on Death. Or, Investing in Life: The History and Future of Pensions*. London/New York (2003).
13. ILO, *Social Protection. A Life cycle continuum investment for social justice, poverty reduction and sustainable development*. Geneva: ILO (2003).
14. R. Maier, W. de Graaf and P. Frericks, Pension Reforms in Europe and Life-course Politics. *Social Policy and Administration* **41** (5), 487–504 (2007).
15. G. Esping-Andersen, *The Three Worlds of Welfare Capitalism*. Princeton: Princeton University Press (1990).
16. H. Arendt, *Vita activa oder vom täglichen Leben*. München: Piper (2002).

17. F. Ewald, *L'Etat Providence*. Paris (1986).

18. M. Kohli, Social Organization and Subjective Construction of the Life Course. In: A.B. Sörensen, F.E. Weinert and L.R. Sherrod (Eds.), *Human Development and the Life Course: Multidisciplinary Perspectives*. Hillsdale, NJ: Lawrence Erlbaum, 271–292 (1986).

19. J. Lewis and S. Guillari, The adult worker model family, gender equality and care: the search for new policy principles and the possibilities and problems of a capabilities approach. *Economy and Society* **34** (1), 76–104 (2005).

20. H. Land, The Family Wage. *Feminist Review* **6**, 55–77 (1980).

21. G. Bonoli, The politics of the new social policies: providing coverage against new social risks in mature welfare states. *Policy & Politics* **33** (3), 431–449 (2005).

22. I. Dingeldey, European Tax Systems and their Impact on Family Employment Patterns. *Journal of Social Policy* **30** (4), 653–672 (2001).

23. G. Esping-Andersen, A New Gender Contract. In: G. Esping-Andersen, A. Hemerijck, J. Myles and D. Gallie, *Why we need a New Welfare State*. Oxford: Oxford University Press, 68–95 (2002).

24. N. Fraser, After the Family Wage. A Postindustrial Thought Experiment. In: N. Fraser (Ed.), *Justice Interruptus: Critical Reflections on the 'Postsocialist' Condition*. New York: Routeledge, 41–66 (1997).

25. S. Lessenich and M. Möhring-Hesse, *Ein neues Leitbild für den Sozialstaat. Eine Expertise im Auftrag der Otto Brenner Stiftung und auf Initiative ihres wissenschaftlichen Gesprächskreises*. Berlin: Otto Brenner Stiftung (2004).

26. B. Jessop, *The Future of the Capitalist State*. Cambridge: Polity Press (2002).

27. P. Frericks, R. Maier and W. de Graaf, Shifting the Pension Mix: Consequences for Dutch and Danish Women. *Social Policy and Administration* **40** (5), 475–492 (2006).

28. Eurostat, Social Protection expenditure in the EU. *News release* **133** (2005).

29. H.-W. Sinn, Why a Funded Pension System is Useful and Why It is Not Useful. *International Tax and Public Finance* **7**, 389–410 (2000).

30. K. Rowlingson, Private Pension Planning: The Rhetoric of Responsibility, The Reality of Insecurity. *Journal of Social Policy* **13** (4), 623–642 (2002).

31. M. Ferrera and M. Rhodes, Building a Sustainable Welfare State. *West European Politics* **23** (2), 257–282 (2000).

32. M. Ferrera, European Integration and National Social Citizenship: Changing Boundaries, New Structuring? *Comparative Political Studies* **36**, 611–652 (2003).

33. E. Durkheim, *The division of labor in society*. Glencoe: Free Press (1964).

34. J. Myles, A New Social Contract for the Elderly? In: G. Esping-Andersen, A. Hemerijck, J. Myles and D. Gallie (Eds.), *Why we need a New Welfare State*. Oxford: University Press, 130–172 (2002).

35. H. Joshi, The Cash Opportunity Costs of Childbearing: An Approach to Estimation using British Data. *Population Studies* **44** (1), 41–60 (1990).

36. N. Folbre, *Who pays for the kids? Gender and the structures of constraint*. New York (1994).

37. T. Knijn and M. Kremer, Gender and the Caring Dimension of Welfare States: Toward Inclusive Citizenship. *Social Politics* **4**, 328–361 (1997).

38. A. Sen, *On Ethics and Economics.* Malden: Blackwell (1988).

39. Y. Shionoya, Economy and Morality. *The Philosophy of the Welfare State.* Cheltenham: Edward Elgar (2005).

40. A. Lipietz, *Towards a New Economic Order. Postfordism, Ecology and Democracy.* Oxford (1992).

41. P. Frericks and R. Maier, The gender pension gap: Effects of norms and reform policies. In: C. Arza & M. Kohli (Eds.), *The Political Economy of Pensions: Politics, Policy Models and Outcomes in Europe.* New York: Routledge, 175–195 (2008).

42. G. Herderscheê, Het pensioenreglement is oneerlijk. *De Volkskrant* 7.1.2004 (2004).

43. M. Veil, *Altersicherung von Frauen in Deutschland und Frankreich. Reformperspektiven und Reformblockaden.* Berlin (2002).

44. Missoc, *Social Protection in the Member States of the European Union, of the European Economic Area and in Switzerland.* Brussels: European Commission (2004).

45. R. Musgrave and P. Musgrave, *Public Finance in Theory and Practice.* New York: McGraw–Hill (1989).

46. J. Hills, *Inequality and the State.* Oxford: UP (2004).

47. A. Kessler-Harris, In Pursuit of Social Citizenship. *Social Politics* **10** (2), 157–175 (2003).

48. N. Fraser and A. Honneth, *Redistribution or Recognition?: A Political–Philosophical Exchange.* London: Verso (2003).

IMAGINATION AND EMPATHY AS CONDITIONS FOR INTERPERSONAL UNDERSTANDING IN THE CONTEXT OF A FACILITATING WORLDVIEW

HANS ALMA

Department of Psychology
University for Humanistics,
Utrecht, The Netherlands
E-mail: h.alma@uvh.nl

ADRI SMALING

Department of Theory of Science and Methodology,
University for Humanistics,
Utrecht, The Netherlands
E-mail: a.smaling@uvh.nl

Some philosophers of science and social scientists disapprove of using empathy in human inquiry. Empathy would be neither sufficient nor necessary for understanding another person. In this article the insufficiency of empathy will be recognized, but the necessity of empathy for interpersonal understanding in everyday life, situations of professional care, and certain forms of human inquiry, especially qualitative research, will be supported. However, empathic understanding should not be conceived as pure psychic identification, but rather as putting oneself imaginatively into the experiential world of another person with the aim to understand the other. Emotional resonance is not only an early phase of it, but also a basic facet. A comprehensive conceptualization of empathic understanding is developed which makes empathy more worthwhile in professional contexts. Empathic understanding is conceived as a two-dimensional concept. The mental dimension refers to affective, cognitive and interpretive facets or phases of empathic understanding and the social dimension refers to expressive, responsive and interactive facets or phases of empathic understanding. These two dimensions are crosswise combined and the most optimal form of empathic understanding is called 'dialogical-hermeneutical empathic understanding'. Furthermore, the importance of imagination and the development of it for optimal empathic understanding is elaborated, as well as its relationship with a worldview.

1. Introduction

In contemporary Dutch society, it is politically correct to speak of multiculturality as a project that was a total failure. The ideal of different cultures and worldviews contributing to a dynamic and open society is deemed naïve and superficial. Integration has become the dominant perspective on dealing with cultural and religious differences instead. This means that people should adjust themselves as soon as possible to the mainstream liberal, enlightened worldview of our secular country. Of course, this development has to do with complex social, economical and political processes which can't be discussed in this article. It is not our purpose to uncritically defend the project of a multicultural society as opposed to processes of integration. We do want to point out, however, that the dominant worldview carries risks with regard to our understanding of and attitude towards other worldviews. Both in science, political debate and social practices the way in which dominant perspectives are taken for granted obstructs a deeper understanding of other cultures and worldviews (especially Islam), the development of new ways for acquiring knowledge and developments in the individual and collective search for finding meaning in life.

In this article, we will defend the view that both for (scientifically) understanding a pluriform society and for finding meaning in it, an attitude is needed that does justice to this pluriformity without losing one's own perspective and orientation. We will argue that professionals — both scientists and spiritual counselors — need an optimal form of empathy. Some philosophers and methodologists do not attach much importance to empathic understanding in social and human research. Some of them are even of the opinion that empathic understanding is dispensable. In contrast with these scholars we will argue that empathic understanding, provided that it is well conceptualized, is indispensable in everyday life, professional care and human inquiry. We do so by undermining their criticisms, sustaining the appraisal of others and developing a concept of optimal empathic understanding in which the development of imaginative competence turns out to be a crucial factor.

2. Criticisms on Empathic Understanding

Some philosophers and methodologists reject empathic understanding as a methodological idea in the social or human sciences. In his textbook *Contemporary Philosophy of Social Science* Brian Fay holds the opinion that empathy is neither sufficient nor necessary for understanding another person.[1] However, he conceives 'empathy' very narrowly and one-sidedly as 'to be one' or 'psychic identification'.

In the opinion of Fay, empathy would not be sufficient for understanding another person because 'to be one' implies 'to have the same or similar experiences as one' and this does not imply understanding these experiences, because understanding an experience does mean more than just having it. Understanding also implies description, interpretation and explanation (p. 18 ff.). Moreover, as far as 'to be one' would also imply 'to have the self-understanding of one', 'to be one' will still be insufficient for understanding because self-knowledge, self-understanding is limited. Self-knowledge and self-understanding are limited regarding one's drives, emotions, motivations, abilities, communicative interactions and mutual influences. In addition, self-knowledge and self-understanding are limited with respect to influences of history, society, institutional contexts, etc. In short, in the opinion of Fay, empathy is not sufficient for understanding one, because, first, having the same experiences as one does not imply understanding (describing, interpreting and explaining) these experiences and, second, self-understanding is limited. Furthermore, according to Fay, empathy would even be unnecessary for understanding another person. His main argument is that others may understand us better than we understand ourselves, because they are not us, they are different from us and do not become us.

Fay's conception of empathy as psychic identification is the basis of his argumentation. We do not agree with this conception, because it is one-sided and too narrow, as will be shown. Consequently, we reject his argumentation. We agree with his conclusion that empathy is not sufficient to understand another in all respects, but we disagree with his conclusion that empathy is not necessary for understanding another person.

2.1. *Empirical-Analytical Philosophy*

Popper, in *Objective knowledge*,[2] conceives empathic understanding a little bit less restrictive than Fay, but he nevertheless rejects it. Popper wants to contribute to a so-called 'objective hermeneutics' by his 'situational analysis' and he rejects "subjective procedures as sympathic understanding or empathy, or there-enactment of other people's actions, or the attempt to put ourselves into another person's situation by making his aims and his problems our own" (Ref. 2, p. 163). Popper does not say that empathy implies psychic identification, but empathic understanding in the sense of 'putting oneself in another's place by making his aims and his problems our own' is still unacceptable to Popper. He does not even discuss a heuristic function. Besides, he still conceives empathic understanding too narrowly by saying that it implies making the aims and problems of another our own.

Other empirical-analytical philosophers, who are logical-positivists or at least strongly influenced by logical positivism, are more tolerant than the critical rationalist Popper. Hempel seems to agree with Fay when he says: "In fact, the occurrence of an empathic state in the interpreter is neither a necessary nor a sufficient condition or sound interpretation or understanding in the scientific sense (...)" (Ref. 3, p. 161–162). (...) "Weber himself stresses that verification of subjective interpretation is always indispensable". However, he also says: "The method of empathic understanding (...) does not in itself constitute an explanation; it rather is essentially a heuristic device" (Ref. 3, p. 239). "This understanding of another person in terms of one's own psychological functioning may prove a useful heuristic device in the search for general psychological principles which might provide a theoretical explanation" (Ref. 3, p. 257–258). So, unlike Popper, Hempel acknowledges a useful heuristic function of empathic understanding, although it would still be dispensable. However, he restricts empathic understanding to empathic identification.

The opinions of Nagel, Rudner and Ryan, in Refs. 4, 5 and 6, and other empirical-analytical philosophers of science, are similar to the opinions of Hempel in so far they acknowledge the heuristic significance of 'Verstehen' or empathic understanding, but deny its explanatory or validational value. For instance, the reasoning of Rudner that empathic understanding is not an indispensable methodological device of the social sciences, runs as follows:

> "The issue is not whether achieving empathic understanding of some subject of inquiry (presumably by an imaginative act of psychologically 'putting oneself into the place of' the subject) is a helpful, fruitful, or indispensable technique for discovering hypotheses or means for testing hypotheses. The issue is not even whether such techniques of discovery are peculiarly techniques of the social scientist. What is at issue is whether empathic understanding constitutes an indispensable method for the validation of hypotheses about social phenomena." (p. 72)

His crucial question is: "What check does the empathizer have on whether his empathic state is veridical (i.e., reliable)?" (p. 73). Rudner claims that if one would have sufficient knowledge to establish independently the reliability of an act of empathic understanding this empathic act would have been made redundant. In case there would not be such knowledge, this act of empathic understanding would appear to be not reliable. Consequently, empathic understanding, even in a broader sense than psychic identification, is needless. However, we think that Rudners requirement that an act of empathic understanding should be checked and sustained in all respects is not reasonable. Indeed, scientific theories or hypotheses, interpretations or even descriptions are rarely exhaustively checked, sustained or verified in all respects. In addition, empathic understanding should be conceived as inherently including a testing procedure to make improvement possible. The inherent testing procedure of our dialogical-interpretive conception of empathic understanding provides it with an explanatory and validational potential. Consequently, empathic understanding can be tested and supported in a reasonable way, without rendering it superfluous.

Furthermore, it will be noticed that these philosophers differentiate sharply between a context of discovery, heuristics or exploration, which would be epistemologically irrelevant, on the one hand, and a context of justification, validation or testing, which would be exclusively epistemologically relevant, on the other hand. Nowadays, since the Kuhnian revolution in philosophy of science, such an exclusive distinction as well as the connected disapproval of the context of discovery as epistemologically meaningless, is out of date (cf. for instance, Ref. 7). Anyhow, Nagel and others acknowledge the relevance or even indispensability of empathic understanding for developing social scientific knowledge. In other words, they acknowledge the methodological significance of empathic understanding. Their far too restrictive view of philosophy of science does not change this.

Not all mentioned empirical-analytical philosophers of science conceive empathic understanding in the very narrow sense of psychic identification. In fact, Rudner and others conceive empathic understanding not as psychic identification, but as 'putting oneself into the place of the subject'. However, all of them underestimate the scientific importance of the context of discovery, the practice of doing research, in which most of them acknowledge the heuristic importance or even necessity of empathic understanding. It is nevertheless interesting that the mentioned logical positivists and their followers are far more tolerant and even appreciating concerning empathic understanding than the critical rationalist Popper.

104

2.2. *Hermeneutic Philosophy*

It might be amazing, that some hermeneutical philosophers, not subscribing to an empirical-analytical approach at all, also reject empathic understanding like the logical positivists or even like Popper. Gadamer,[8] rejects understanding in the sense of re-enactment, an empathic re-experience or reconstruction of an original intention, a recovery of the (past) intentions of agents, an identification with or a re-creation of the minds of agents. Interpretation is not a psychological process of empathy, not the empathic re-experience of an original intention. In this respect Gadamer rejects the views of Schleiermacher, the young Dilthey, Collingwood and other hermeneutical philosophers from the Romanticist tradition, which particularly dominated in the nineteenth century. Within this tradition 'Verstehen' (understanding) has been mainly conceived as empathic understanding in the sense of re-enactment, recovery, identifying and grasping the authentic and unique feelings and experiences of another. By the way, Gadamer uses the term empathy ('Einfühlung') only a few times. More often he uses terms like 'psychological interpretation' or 'sympathetic understanding' ('Mitleidenschaft'; a kind of loving relationship), mainly when he discusses the hermeneutics of Schleiermacher. In Gadamer's view, interpretive understanding is rather a 'fusion of horizons', placing other's behavior or its products in a historical, social-cultural tradition. Historical figures, authors and artists are not conscious of the workings of these traditions. However, Gadamer did neither claim to develop or describe a methodology for the social or human sciences nor to present a psychotherapeutic theory or approach. His book *Truth and Method* (1975; originally in German, 1960)[8] is about hermeneutical philosophy and the philosophy of the human sciences ('Geisteswissenschaften'), which does not simply imply a specific scientific or therapeutic methodology. Hermeneutical explanation in the sense of Gadamer focuses on culture and history and abstracts from several aspects of human, societal and personal everyday life.

Sometimes Gadamer seems not to reject empathic understanding in all respects or possible meaning aspects of it. He says:

> "When we try to understand a text, we do not place ourselves in the author's inner state; rather, if one wants to speak of 'placing oneself', we place ourselves in his point of view. But this means nothing else than that we try to let stand the claim to correctness of what the other person says. We will even, if we want to understand, attempt to strengthen his arguments." (Ref. 9, p. 69)

However, he adds also: "We can (...) leave completely aside what Schleiermacher set forth as subjective interpretation." And: "It is the task of hermeneutics to illuminate this miracle of understanding, which is not a mysterious communication of souls, but rather a participation in shared meaning" (p. 69). Thus he rejects empathic understanding in a more narrow sense. He rejects especially the emotional and non-rational dimensions of understanding.

Unlike Gadamer, Ricoeur explicitly discusses a hermeneutical methodology of the human sciences in his book *Hermeneutics and the human sciences*.[10] In chapter eight 'The model of the text: meaningful action considered as a text' he defends the text as a paradigm for the object of the social sciences as well as the methodology of text-interpretation as a paradigm for interpretation in general in the field of the human sciences. Although Ricoeur does not use the terms 'empathy' nor 'empathic understanding', he rejects, like Gadamer, empathic understanding by implication in so far it would mean:

> "an immediate grasping of a foreign psychic life or with an emotional identification with a mental intention. Understanding is entirely mediated by the whole of explanatory procedures which precede it and accompany it. The counterpart of this personal appropriation is not something which can be felt, it is the dynamic meaning released by the explanation which we identified earlier with the reference of the text, i.e., its power of disclosing a world."
> (Ref. 10, p. 220)

Because Ricoeur does not use the term empathy nor empathic understanding it is not sure what his opinion concerning a much broader conception of empathic understanding would be. Anyhow, Ricoeur restricts his model to theoretical interpretation and explanation in the human sciences and he abstracts, like Gadamer, from the more practical methodological domain of collecting data, such as interviewing, participatory observation, participative inquiry, etc. However, we wish to stress that precisely in this practical domain of doing empirical research empathic understanding, in a broader sense than psychic or emotional identification, is not dispensable. It seems to be that Gadamer and Ricoeur, like the mentioned empirical-analytical philosophers, neglect or underestimate the importance of the context of discovery, the methodological domain, the practice of doing empirical research. In addition, social science in the opinion of Gadamer and Ricoeur abstracts from explaining individual differences in terms of purposeful or intentional

106

actions. Their conception of social science (psychology included) turns out to be too restrictive.

Anyway, Gadamer as well as Ricoeur abstract from the fact that the classical idea of understanding as trying to grasp, determine or interpret the mental state or mental process (intentions, motivations, thinking, feeling, etc.) of another person is indispensable for everyday social life, doing human research such as interviewing, participatory obervation and participative action research, and other professional practices such as education, nursing, mental care, psychotherapy and counseling. Our conceptualization of empathic understanding will incorporate these issues.

We also disagree with methodologists of the social or human sciences who reject the necessity of empathy, conceive it very narrowly or one-sidedly, or disapprove the methodological worth of empathy, sometimes without mentioning the term, such as Alvesson, Seale and Silverman in Refs. 11, 12 and 13. For instance, Seale says that the interpretivist and feminist approaches in social research "can be seen to be somewhat romantic, believing that authentic accounts of what 'things are really like' will be given in moments of emotional intimacy where souls are bared and pretence is stripped away" (Ref. 12, p. 209). "There is a danger here of imagining that a particular interaction format is an automatic guarantee of the analytic status of the data that emerge" (Ref. 12, p. 209). In contrast with him, we will bring to the fore that empathic understanding does not necessarily imply uncritical acceptance of empathic experiences as always having validational or self-evidential value of themselves. Empathic understanding is more than emotional intimacy and it does not simply imply an authentic account of the true soul of another. This will be sustained on the basis of the relevant literature.

3. Appreciations of Empathic Understanding

In the former section different opinions of scholars have been discussed who seem to have a negative stance towards empathic understanding. Some empirical-analytical opinions (of, for instance, Carl Hempel and Karl Popper), hermeneutic-interpretive opinions (of, for instance, Hans-Georg Gadamer and Paul Ricoeur) and methodological opinions (of, for instance, Seale and Silverman) correspond with Fay's radical negative stance, but only partly. Sometimes these opinions imply a broader concept of empathic understanding. Sometimes they abstract from the practical domain of doing empirical research, ignore the importance of the practice of doing empirical research or neglect ordinary life. Sometimes empathic understanding is seen as an indispensable heuristic device for developing social scientific knowledge. In this section different opinions of philosophers, scientists and practitioners are discussed who have a more positive stance towards empathic understanding.

3.1. *Philosophy of Science*

As already said Gadamer and Ricoeur criticize the classical hermeneutic philosophers Friedrich Schleiermacher (philosopher of theological research) and Wilhelm Dilthey (philosopher of historical research), because they conceive the aim of hermeneutics as the reconstruction of the mental state and process of the author of a text. Their objection concerns mainly the psychologistic approach. However, Schleiermacher and Dilthey do not simply conceive hermeneutic understanding as psychological identification. From the beginning Schleiermacher had in mind two basic parts of hermeneutics: objective (grammatical or linguistic) understanding and subjective (psychological) understanding. Both parts should be integrated with each other to reach true understanding (See Ref. 14; see also Ref. 15). The older Schleiermacher accentuated subjective understanding (one must have an understanding of man himself in order to understand what he speaks). The younger Dilthey likewise stressed psychological understanding, but in his later life he abandoned this sheer psychological approach. Understanding could also mean recovering oneself in another, studying other persons to understand one's own culture, for instance (See Ref. 16; and also Ref. 15).

Max Weber, philosopher of social science and originator of interpretive sociology, has proposed the methodological 'principle of subjective interpretation'.[17] This principle states that sociological phenomena have to be studied in terms of the thoughts, feelings, motivations and purposeful behavior (action) of the studied, abstract or concrete, subject. In his opinion, subjective interpretation is based on two kinds of understanding: rational understanding (think of logic, mathematics or statistics; and also of understanding rationally purposeful action) and emotional understanding (sometimes called empathic understanding or emotionally empathic understanding), which is imaginative participation in "such emotional reactions as anxiety, anger, ambition, envy, jealousy, love, enthusiasm, pride, vengefulness, loyalty, devotion, and appetites of all sorts, and thereby understand the irrational conduct which grows out of them." (Ref. 18, p. 23) Sociological interpretation is mainly rational understanding, but empathic understanding is a necessary component. Empathic or emotional understanding is needed for generating hypotheses to be tested. Hence, in Weber's view empathic understanding is not sufficient, but still necessary, or, at least, heuristically important for generating hypotheses, while rational understanding is necessary and insufficient as well. In this respect Max Weber has influenced the phenomenological philosopher and sociological theorist Alfred Schutz.[19,20] Schutz stresses the point that 'subjective interpretation' should not mean empathy with unobservable, unverifiable inner states such as intentional emotions. Schutz's own answer to the charge of subjectivism made by Ernest Nagel and others is unambiguous:

> "(...) a method which would require that the individual scientific observer identify himself with the social agent observed in order to understand the motives of the latter, or a method which would refer the selection of the facts observed and their interpretation to the private value system of the particular observer, would merely lead to an uncontrollable private and subjective image in the mind of this particular student of human affairs, but never to a scientific theory. *But I do not know of any social scientist of stature who ever advocated such a concept of subjectivity*" (our italics).
> (Ref. 19, p. 52)

Fay, Silverman and others should have realized this! By implication, Schutz does not see Max Weber as taking such a subjectivist position. Indeed, according to Max Weber both empathic and rational understanding are necessary ingredients of sociological understanding.

The philosopher of social science Peter Winch, who is deeply influenced by Ludwig Wittgenstein, conceives understanding mainly as understanding rule-following or rule-governed behavior. However, he does not reject empathic understanding in all respects:

> "(...) a historian or sociologist of religion must himself have some religious feeling if he is to make sense of the religious movement he is studying and understand the considerations which govern the lives of its participants. A historian of art must have some aesthetic sense if he is to understand the problems confronting the artists of his period (...)" (Ref. 21, p. 88)

He calls this feeling or sense 'historical imagination' and also 'empathy' (Ref. 21, p. 90). He even defends the idea of 'putting oneself in the other fellow's position' against criticisms of, for instance, Popper. Popper says that hypotheses based on such a mental process must be tested. However, Winch shows that Max Weber himself already insists that 'putting oneself in the other fellow's position' must be tested. Weber requires a process of checking the validity of 'intuitions' that are based on empathy. So, Popper cannot blaim Weber for accepting 'intuitions' on the basis of empathy as sufficient evidence. It is true that Winch is of the opinion that Weber gives a wrong account of this process as a statistical process, but: "Weber is clearly right in pointing out that the obvious interpretation need not be the right one" (Ref. 21, p. 113). To summarize, in the view of Winch, empathic understanding needs a checking process, but the heuristic function of it within understanding rule-governed behavior is indispensable. In our conceptualization of optimal empathic understanding a checking process will be incorporated as well as the feeling ingredient.

3.2. *Phenomenological Philosophy*

Edith Stein shows in her dissertation 'On the problem of empathy' (originally published in 1917)[22] that several philosophers have acknowledged the possibility and epistemological significance of empathy. Her conception of empathy differs from Fay's and shows similarities with the conception of the early Edmund Husserl in his 'Ideas'. (See Ref. 23; originally in German: 'Ideen', 1913.) She says: "Empathy in our strict defined sense as the experience of foreign consciousness can only be the non-primordial experience which announces a primordial one. It is neither the primordial experience nor the 'assumed' one" (p. 14). So, in her opinion and in the opinion of several others empathy is not psychic identification at all. Empathy in the sense of feeling into, 'Einfühlung', 'Nachfühlen', or 'Nacherleben' is not sympathy in the sense of fellow feeling, feeling with, 'Mitfühlung' or 'Mitgefühl'. It is not having or simply sharing the primordial experience of another. The empathic experience is primordial as present experience, but it is non-primordial in content (cf. p. 10). At the same time empathy is neither a pure matter of producing cognitive assumptions. It is not sheer hypothesizing or theorizing. Empathy is a kind of act of perceiving sui generis (cf. p. 11). Empathy differs from sympathy as well as feeling of oneness, because the experience of the empathizer is non-primordial in content, while the experience of the empathizee is primordial in all respects. Empathy also differs from outer perception (including sensory perception) insofar "outer perception is a term for acts in which spatio-temporal concrete being and occurring come to me in embodied givenness" (p. 6). Furthermore, empathy differs from memory, expectation or fantasy, because the subject of the empathized experience (the empathizee) is not the subject empathizing (the empathizer), but another (cf. p. 10). According to Stein, empathy in the indicated sense is an essential aspect of self-understanding as well as understanding others. (To be clear, the concept of 'self empathy' should be understood as implying that one empathically relates to oneself as another!) Stein does not conceive empathy as psychic identification. At the same time she seems to exclude hypothesizing and theorizing from the empathic act at all. We do not follow her in this respect. We understand her position as rather Husserlian and as a consequence not sufficiently hermeneutical. In our conception of empathic understanding, hermeneutics, especially the idea of the hermeneutical circle, will be an important ingredient. Empathic understanding as a process, certainly in human inquiry, should include this hermeneutical circle, which implies testing, adjusting and developing interpretations. Nevertheless, in his transcendental phenomenology Husserl needs the idea of empathy ('Einfühlung', 'Fremderfahrung als Appräsentation oder analogische Apperzeption') to constitute intersubjectivity between the isolated transcendental ego's, to make possible a social world.[24] Thus, in the philosophy of Husserl empathy, be it in a non-hermeneutical sense, is of crucial importance.

Maurice Merleau-Ponty tries to sustain the possibility of understanding another by an epistemological, and later on, an ontological grounding of intersubjectivity on a sort of basic, embodied, pre-subjective commonality, a shared 'intercorporeity'.[25,26] However, he does not discuss empathic understanding as a way of personal handling, as a method which may be improved, developed further and checked on its validational value. Nevertheless, the idea of embodiment of mutual empathic understanding will be incorporated in our conception of empathic understanding. Empathic understanding will be based on emotional resonance which may be grounded in that pre-subjective 'intercorporeity'.

Joachim Duyndam, in Ref. 27, who focuses on the emotional aspects of empathy in psychotherapy and counseling and not so much on attitudinal or behavioral aspects, says that empathy does not imply sharing the actual feelings of another, being absorbed by the actual feelings of another. "Empathy is not the same as collectively bathing or immersing in actual feelings" (p. 144). True empathy implies having one's own potential feelings connected with the actual feelings of another. In his opinion the supporting or empowering effect of empathy should be explained by its potential character (for instance, the realization that 'this could happen to me') and not by actuality (p. 142 ff). (The distinction between potentiality and actuality is borrowed from Husserl. We notice the similarity with Stein: the actual feelings of Duyndam are similar to the primordial experiences of Stein and the potential feelings of Duyndam correspond to the non-primordial content of the empathic experience of Stein.) Striving for having the actual feelings of another may even lead to an untrue and dangerous form of empathy, which Duyndam calls 'gruesome sham-empathy'. In addition to the misunderstanding that empathy means having the same feelings as another Duyndam points to a second, related misunderstanding: that empathy would imply unquestioning support. Empathy does not necessarily imply that the empathizer subscribes to the truth-claims or good-claims of the empathizee concerning the outside world. Duyndam nevertheless recognizes the importance of 'true' or 'good' empathy. In his view a positive therapeutic effect of empathy depends on the potential character of empathy and not on imitating or taking over the actual feelings from the empathizee, the client, by the empathizer, the therapist. At the same time he stresses the emotionality of empathy, be it a potential and not an actual sharing in someone's emotions. Moreover, he stresses the importance of an empathizing other for self-understanding. However, Duyndam seems to underestimate the role of the actual or primordial feelings which the empathizer does experience besides the potential or non-primordial feelings as Stein points out. In other words, he ignores the importance of simulation or resonance of emotions as basic aspects of empathy. As a matter of fact, autistic and psychopathic persons cannot completely compensate for their affective deficiency by learning cognitive or rational strategies. They continue to be non-empathetic or weak-empathetic.

To conclude, the conception of empathy as psychic identification seems at least one-sided and too narrow. Empathic understanding in a broader sense, for instance as 'imaginative participation in emotional reactions with the purpose of understanding irrational conduct which grows out of them' (Weber), 'putting oneself in the other fellow's position' (Winch), 'having a non-primordial experience of foreign consciousness' (Stein) or 'having one's own potential feelings connected with the actual feelings of the other' (Duyndam), does not imply being one, being similar to one or having the same or similar experiences as one. Moreover, these scholars recognize the indispensability of empathic understanding for human communication. In the following, their views will be sustained and elaborated. In our conceptualization we will distinguish empathy from sympathy and explicitly reckon with the danger of sham-empathy and the misunderstanding that empathy would imply subscription, without ignoring the primordial affective component. In addition the importance of social interaction for improving empathic understanding of another will be explicated. The quality of empathic understanding can be developed further by hermeneutic and dialogical processes. Most philosophers discussed above abstract not only from the methodological domain and other professional contexts, but also from the dimension of social interaction and its practical relevance for improving empathic understanding.

3.3. *Psychotherapy and Counseling*

If we conceive empathy as psychic identification and also say that this is impossible or worthless, like Fay, a problem arises concerning psychotherapy, counseling, developmental aid, rearing in early childhood, etc. After all, empathic understanding is generally recognized as of great importance to these domains. However, in these cases empathy is conceived in a broader sense, not in the sense of being one or being very similar to one. Let us have a look at two of the most influential pioneers in the study of empathy in psychotherapy: Carl Rogers and Heinz Kohut. In the context of this article we do not discuss the different opinions on the issue in what way empathy or empathic understanding would be a therapeutic instrument or just a condition and what would be their effectiveness as a therapeutic instrument. To both Rogers and Kohut empathy is at least the most important method to understand another. We focus on the conceptual meaning of empathic understanding, especially with regard to psychic identification. It will be apparent that even these two champions of the use of empathy in therapy do not simply conceptualize empathic understanding as psychic identification.

In client-centered psychotherapy empathic understanding is one of the three most important conditions for personal growth in a therapeutic situation. (See Refs. 28, p. 33 ff.; 29, p. 60 ff.) The other two conditions are genuineness (congruence or transparency) and acceptance of the other as a separate person (unconditional positive regard). Sometimes Rogers adds two other conditions for significant learning in therapy: facing a problem (the client should have an uncertain and ambivalent desire to learn or to change, growing out of a perceived difficulty in meeting life) and, what we would call, responsivity (the client should experience or perceive something of the therapist's acceptance, congruence and empathy). (See Ref. 28, p. 281 ff.) We will come back to this issue of responsivity later on.

Together these necessary conditions could be sufficient to form an effective psychotherapeutic instrument.

In the context of this essay, it is important to notice that Rogers emphasizes the 'as if' nature of empathic understanding:

> "To sense the client's inner world of private personal meanings as if it were your own, but without ever losing the 'as if' quality, this is empathy, and this seems essential to a growth promoting relationship. To sense his confusion or his timidity or his anger or his feeling of being treated unfairly as if it were your own, yet without your own uncertainty or fear or anger or suspicion getting bound up in it, this is the condition I am endeavoring to describe." (...) "I am sure that when the counselor can grasp the moment-to-moment experiencing occurring in the inner world of the client, as the client sees and feels it, without losing the separateness of his own identity in this empathic process, then change is likely to occur. (...) Suitable training experiences have been utilized in the training of counselors (...). Such experiences enable the person to listen more sensitively, to receive more of the subtle meanings the other person is expressing in words, gesture, and posture, to resonate more deeply and freely within himself to the significance of those expressions." (Ref. 29, p. 89, 90; see also Ref. 28, p. 62)

Already in Ref. 30 is written: "(...) the experiencing with the client, the living of his attitudes, is not in terms of emotional identification on the counselor's part (...)" (p. 29). Rogers acknowledges that empathic understanding is neither simply psychic identification, being one, or being similar to one (cf. Ref. 31), nor "a wooden technique of pseudo-understanding in which the counselor 'reflects back what the client has just said' ". (See Ref. 29, p. 90 ff.) Likewise, Rogerian empathic understanding is neither pure simulation of the thoughts and feelings of another nor mere theorizing. Affective and cognitive aspects of the empathic attitude are integrated in empathic understanding, which can be trained in interactions.

The 'as if' quality of empathy does not make it 'sham-empathy'. On the contrary, empathy in the sense of Rogers means neither being absorbed nor being totally immersed in actual or primordial feelings. At the same time this 'as if' quality does not preclude empathy to be genuine. The 'as if' quality refers to the interpretive or critical dimension of empathic understanding. (One may also think of the non-primordial and potential dimensions of empathy as discussed in a former section.) Bozarth in

Ref. 32 says: "Whatever the reason, he (Rogers) seemed particularly concerned that the therapist not identify with the client but maintain the as-if dimension" (p. 86). Bozarth mentions this concern as an aspect of Roger's idea of objectivity in psychotherapy. Furthermore, we notice that to Rogers empathic understanding is not a pure cognitive-rational activity. Empathy is basically a highly sensitive and open attitude, which, in addition, needs to be expressed and communicated to be effective and, more important in our context, also to be checked. The behavioral expression to the other of empathic understanding may help to improve this understanding.

Eugene Gendlin, who has developed Rogerian client-centered therapy further, makes the therapist-client distinction even greater. The therapist has to use and analyze his own experiences for empathically understanding the emotions and feelings of the client. Gendlin calls this experiential aspect of empathy 'the experiential response'.[33] The understanding of another's consciousness by the aid of your own has been called 'sympathic introspection' by the sociological methodologist Cooley.[34] We do not embrace his term. 'Empathic introspection' would be a better one. For empathy does not necessarily imply sympathy. Empathy is more neutral than sympathy, which implies appreciation and a certain degree of commitment.

Denijs Bru, psychotherapist, humanistic counselor and inspired by Rogers and Gendlin, stresses the importance of grasping empathy as an interactional process. He is of the opinion that empathy is more than an attitude. Empathy starts as a sensitive attitude, but continues as an interactional process, which aims at mutual understanding. Further, this mutual understanding is more than communication or responsivity, because the client as well as the therapist or counselor may change or develop as a result of it (Ref. 35, p. 52). The important social-behavioral dimension will be integrated in our conceptualization of empathic understanding.

The psycho-analytical therapist and theorist Heinz Kohut defines empathy as 'the capacity to think and feel oneself into the inner life of another person' (Ref. 36, p. 82). In his opinion empathy is the chief method of understanding in therapeutical practice as well as theoretical investigation. Empathy is not a form of extra-sensory perception. Nor is it the same as we would feel if in similar circumstances. Projection in this sense is not empathy. Empathy is also not the same as identifying with, or becoming the other, so that one is flooded by or overwhelmed by the intensity of another's feelings. On the contrary, empathy implies trial and error, a long-term process of 'tasting' and checking. Kohut also characterizes empathy as 'vicarious introspection', without pretending identification with the other.[37,38]

By 'vicarious introspection' he means 'that only through introspection in our own experience can we learn what it might be like for another person in a similar psychological circumstance' (Ref. 39, p. 247). "For Kohut, empathy is that which allows an individual to experience another's experience without losing one's ability to evaluate objectively another's mental states. In other words, empathy is simply experience-near observation and nothing more" (Ref. 39, p. 248). So, like Rogers, Kohut conceives empathy as implying a kind of objectivity.

The psychiatrist and poet Van den Hoofdakker, in Ref. 40, is of the opinion that empathy is a necessary condition for an effective and humane medical practice, but that empathy is not intrinsically good in an ethical sense. Empathic understanding may be used to exercise power, to manipulate or to hurt another. Indeed, empathic understanding implies knowing the weaknesses of another. A smart manipulator needs empathic understanding. To be sure, in such a case empathy has nothing to do with being one. The psychiatrist Van Tilburg, in Ref. 41, accentuates the importance of training courses in medical education to promote the empathic abilities of physicians. In addition, Bohart and Greenberg, in Ref. 42, show in their volume 'Empathy Reconsidered: New Directions in Psychotherapy' (p. 4 ff.) that empathy is not sheer psychic identification and that aspects like feeling, thinking, interpreting, communication, co-construction, responsiveness and behavioral achievement play a part.

In sum, in psychotherapy empathy does not imply psychic identification. Even the two champions of using empathy, Rogers and Kohut, warn against striving for identification! Empathic understanding means using one's sensitivity, but without losing the 'as if' quality and a kind of 'objectivity'. In other words, Rogers, Kohut and others, recognize the necessity of the affective or resonant facet of empathy, but they stress just as much the necessity of the cognitive or rational facets of empathy: checking, testing and analyzing interpretations. Further, empathy is an attitude or a capacity to think and feel oneself into the experiences of another, but, in addition, the process of empathic understanding needs the social dimension, the interaction between empathizer and empathizee. This social dimension becomes an important factor, because it enhances the quality of empathic understanding. Moreover, this interactional empathic process is not only a checking and correcting process, but may also change the participating persons. Nevertheless, empathic understanding is not necessarily 'good' or an ethical principle. First of all, empathic understanding is an indispensable way to understand others. Besides, explicit training programs for enforcing the empathic capacity should be included in medical education. These elements of empathic understanding will be incorporated in our conceptualization.

3.4. *Methodology of Social Scientific Research*

If we conceive empathic understanding as narrowly and one-sidedly as Fay in Ref. 1 does, several methodological concepts would turn out to be problematic. Some well-known concepts are: 'role taking' or 'taking the perspective of another', 'getting good rapport', and 'open-mindedness'. Should we reject these concepts like Fay rejects empathic understanding? All these concepts overlap a great deal in meaning and are strongly connected with empathy! Or should we reject the general methodological significance of empathy and the other mentioned concepts because they would be exclusively linked with outdated and dangerous Romanticism as Silverman in Ref. 13 and Alvesson in Ref. 11 do? No! The main reason is that empathy in connection with these methodological concepts is not meant as psychic identification, being one or being like one, nor as a Romanticist infallible entrancement into the authentic and unique subjective experiences of another.

For instance, Rubin and Rubin say in their book on qualitative interviewing that throughout the research interview the interviewer should show that he or she empathizes with the present emotional undertones.[43] And they discuss some ways to do that. However, empathic understanding "does not require becoming a member of the group you are studying. You can study trophy hunters without being one. (...) Disagreement does not necessarily conflict with empathy and rapport in fieldwork" (Ref. 43, p. 133). Moreover, Kvale says in his book on qualitative research interviewing: "A good interviewer is an expert in the topic of the interview as well as in human interaction" (Ref. 44, p. 147). And:

> "The interviewer has an empathic access to the world of the interviewee; the interviewee's lived meanings may be immediately accessible in the situation, communicated not only by words, but by tone of voice, expressions, and gestures in the natural flow of a conversation. The research interviewer uses him- or herself as a research instrument, drawing upon an implicit bodily and emotional mode of knowing that allows a privileged access to the subject's lived world." (Ref. 44, p. 125)

Kvale outlines ten criteria for interview qualifications that may lead to good interviews. His fifth criterion (The interviewer should be sensitive!) runs as follows:

> "The interviewer listens actively to the content of what is said, hears the many nuances of meaning described more fully. The interviewer is empathic, listens to the emotional message in what is said, not only hearing what is said but also how it is said, and notices as well what is not said. The interviewer feels when a topic is too emotional to pursue in the interview." (Ref. 44, p. 149)

Thus, to these authors on qualitative interviewing empathic understanding is not psychic identification, being one or being like one. Rather empathy is a very sensitive mode of interpretive listening which should be shown.

Even Gordon emphasizes the importance of empathy in his conventional book on interviewing, not specifically qualitative interviewing.[45] He defines empathy as:

> "the process by which one person is able to imaginatively place himself in another's role and situation in order to understand the other's feelings, point of view, attitudes, and tendencies to act in a given situation. In essence, empathy is the ability to correctly answer the question, 'How would I feel or act in the situation if I were in his place?'." (Ref. 45, p. 13)

In this conception empathic understanding can easily be seen as a type of taking the role (or perspective) of another. In addition, Gorden mentions three conditions for a person's ability to successfully empathize with another person in a situation:

(1) The degree to which this person's knowledge of the other's situation is complete and accurate.
(2) The extent to which this person has experienced the same situation, or the degree to which he can imaginatively construct such a situation from elements of several similar situations.
(3) The degree to which this person accurately observes and remembers his own experiences.

The emotional and affective components of empathy are somewhat suppressed in the formulation of these conditions. Indeed, the cognitive components are emphasized, including the ability to use one's own experiences to interpret the meaning of what another is saying (we recognize this element

118

of using one's own experiences in the views of Bru, Duyndam, Gendlin, Cooley, Kohut and others). Again, empathic understanding is clearly different from psychic identification, pure emotional indwelling as well as Romanticist mergence. The conceptions of empathic understanding which Alvesson, Fay, Seale and Silverman seem to endorse are too narrow and one-sided. This is the main reason why their critical comments regarding the epistemological and methodological significance of empathic understanding in human inquiry turn out to be irrelevant.

4. Towards a Conception of Optimal Empathic Understanding

Fay, in Ref. 1, says that empathic understanding is insufficient as well as unnecessary for understanding another person. We disagree. The main reason turned out to be that Fay conceives empathic understanding too narrowly and one-sidedly. Conceived in a broader sense empathic understanding has been indicated by several philosophers, scientists, therapists and counselors as indispensable for human understanding. For instance, in addition to the former texts, the indispensability of empathic understanding in ordinary life is shown in developmental psychology.[46] Empathic understanding has to develop on the basis of a kind of proto-empathy, a type of emotional immersion, a bodily felt resonance with the mother. Neuroscience has shown that a neurophysiological mechanism is necessary for the development of the capacity of empathic understanding: the system of the 'mirror-neurons'. Malfunctioning of these 'mirror-neurons' may have autism (no empathic competence) as a consequence.[47–49] In cognitive science the necessity of empathic understanding in everyday life is also acknowledged. Cognitive scientists differentiate between three aspects or phases of empathy in everyday life: simulation (pre-linguistic and biologically based simulation of the emotions of another), theorising (using 'folk-psychology', or primitive theories of mind) to make sense of another's emotions and behavior, and analogical reasoning (you perceive another's behavior that you know of yourself and you project your own remembered emotions which are associated with your own behavior on the other). (See for instance Refs. 50 and 51; and see for a more elaborated version of the underpinnings of the following conceptualization, Ref. 52.)

Which conception should we choose? If empathic understanding is neither psychic identification, being one, nor being similar to one, then what is it? Based on the views discussed before, we propose the following definition of 'empathy'.

Empathy is the ability of placing oneself imaginatively in another's experiential world while feeling into her or his experiences (points of view, thoughts, ideas, cognitions, desires, intentions to act, and, especially, motivations, feelings and emotions).

The following should be kept in mind:

(1) empathy does not mean psychic identification, being one or being similar to one; it has an 'as if' quality without becoming a sort of sham-empathy;

(2) empathy is not just projecting what we would feel in a situation similar to the situation of another; it is a trial and error, long-term, 'tasting' and checking process;

(3) empathy does not mean 'feeling with' ('Mitfühlung', fellow feeling or sympathy), but it means 'feeling into' ('Einfühlung', 'Nachfühlung' or 'Nacherleben') the experiences of another; nevertheless, a sort of emotional resonance is basic;

(4) the experiential world of the other is seen as having cognitive, affective, emotional and motivational dimensions;

(5) empathy as such is an attitudinal ability and not yet communicative (verbal or non-verbal) behavior; however, in the process of empathic understanding empathy may get a communicative character;

(6) empathy does not necessarily imply subscription to what is expressed; the empathizer may disagree with the empathizee;

(7) empathy neither implies ethical goodness; empathy may be used to exercise power, to manipulate and to hurt another;

(8) the empathic ability has to be developed through one's personal life; its quality is age-related; so, the development of empathy may have certain stages.

Having defined 'empathy' we can define 'empathic understanding' as follows:

Empathic understanding is understanding another person on the basis of empathy; this understanding is directed at comprehending or explaining the experiences, mental states and behavior of that person, also in their interrelationships.

Or, more shortly:

> *Empathic understanding is placing oneself imaginatively in an-*
> *other's experiential world while feeling into her or his experiences*
> *with the aim of comprehending these experiences.*

Comprehension does not imply sympathy or approval; it is a type of interpretive explanation. Neither empathy nor empathic understanding imply actually experiencing the mental state of the other or showing the behavior of the other. Although empathic understanding will still be insufficient for complete understanding of another person in all respects, it will be necessary and feasible. Admittedly, this conception expresses why understanding another is difficult and may fail, but it also expresses that empathic understanding is not a hopeless endeavor. Empathic understanding is possible. It is not a mystery. It is not white magic. As a matter of fact, empathic understanding is a necessary condition for understanding others in ordinary life, for keeping personal mental health and for maintaining human societies. Moreover, during a lifetime or during learning processes empathic understanding can be improved within communicative processes. Because optimizing the development of empathic understanding needs communication, a conceptualization of optimal empathic understanding must include a socially behavioral dimension in addition to the dimension of attitudinal or mental ability.

Van Strien,[53] like Gendlin and Bru,[33,35] does not abstract from the dimension of social behavior, especially social interaction. He differentiates between five phases or facets of the process of empathic understanding of another person. Van Strien writes within the context of psycho-analytic therapy. We will adapt his descriptions and characterizations to the research context and eliminate the typical psycho-analytic interpretations:

- *affective empathy* or *empathic resonance*; the empathizer participates, in an 'as if' mode, in the experiential world of another person; simulation may be the case on this level;
- *cognitive empathy*; the empathizer analyses and interprets the perceived affective experiences and behaviors of another; using a primitive theory of mind may be the case;
- *expressed empathy*; the empathizer expresses his or her experienced empathy, verbally or nonverbally; in an interview situation empathic understanding should be shown to the interviewee to have an effect; and also in a situation of giving (health) care to express empathy may be an essential contribution to the effectiveness of care giving;
- *received empathy*; the empathizee has to accept the expressed empa-

thy; this responsivity is necessary to get good rapport between the researcher and the researched; so, we could also speak of responsive empathy; and in a situation of giving (health) care responsivity may improve the effectiveness; in addition, one may say that care is only true care if the caring activities are accepted by the care receiver (see Ref. 54);

- *interactional empathy*; the empathizer and the empathizee interact; they react appropriately to each other, especially concerning the aspect of empathic understanding; in the interactive process empathic understanding is expressed, received, accepted, affirmed and stimulated.

The affective and cognitive types of empathy are of a mental or attitudinal nature and the other three types are of a socially behavioral or communicative nature.

We want to modify and extend this series from Van Strien in Ref. 53 in four respects.

First, we add a third, hermeneutical-interpretive type (in short interpretive empathy) to the mental types. This interpretive type of empathy has already been discussed in former sections:

- *interpretive empathy*; a hermeneutical approach to empathic understanding implies that the verbal and non-verbal behavior of the interpreting empathizer has to be understood and interpreted by means of a process characterized by the so-called hermeneutical circle; in the hermeneutical circle parts of the behavior of the researched empathizee are interpreted and re-interpreted from the whole, and the whole is interpreted and re-interpreted from the parts; this understanding and interpreting process is a checking and validating process; this interpreting process also implies self-clarification and self-understanding on the part of the interpreting empathizer because his or her for-understandings or presuppositions are being confronted by the behavior of the researched; in addition, societal, cultural and historical aspects are included in the interpretive process. In addition, the other is also interpreting his or her own behavior, thoughts, feelings, etc. So, understanding another empathically implies a 'double hermeneutics' (cf. Ref. 55); besides, the social dimension of empathic understanding also stimulates the empathizers's self-clarification, self-understanding and self-development.

Secondly, in our opinion, the set of mental types of empathy (affective, cognitive, interpretive empathy) and the set of social types of empathy (expressed, received, interactional empathy) can be combined, because these two sets can be seen as independent dimensions; we can construct a cross product of these sets or dimensions (see Figure 1). The mental types may exist without a social dimension, but they need a social dimension as indicated to be optimized as well as to be made more effective or worthwhile in professional situations such as doing interviews and giving health care. However, the social types cannot exist without a mental dimension. Therefore, the social dimension has also a value 'not expressed', a kind of zero-value. In situations of professional care giving, counseling, education, research-interviewing and participative inquiry the mental as well as the social dimension of empathic understanding should be optimized. Explicit training courses will be indicative. Hence, the 3x4 cross table in Figure 1 represents a typology of twelve types of empathic understanding of which the dialogical-hermeneutical type is the most optimal one.

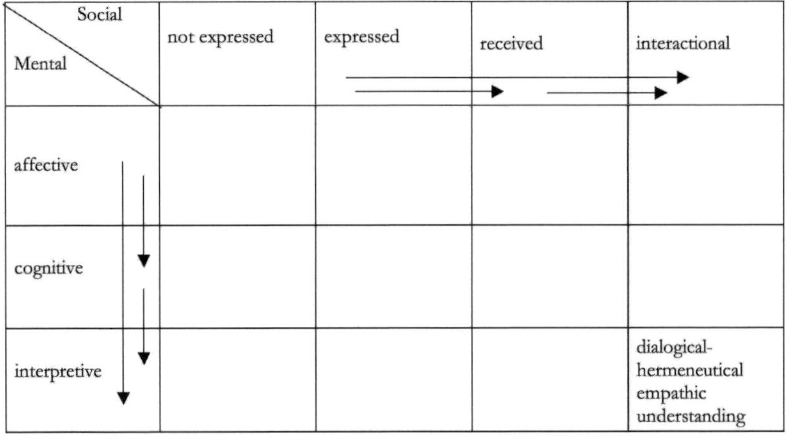

Social Mental	not expressed	expressed	received	interactional
affective				
cognitive				
interpretive				dialogical-hermeneutical empathic understanding

Figure 1. A 3x4 structural conceptualization of empathic understanding.

Thirdly, 'expressed empathic understanding' should be read as 'empathic understanding expressed to another', while the 'zero-value' of the social dimension of empathic understanding concerns empathic understanding which is not expressed to another. In this way, two variations within this category of 'zero-value' can be differentiated: (i) empathic understanding which is purely mental, not expressed at all, and (ii) empathic understanding which is expressed but not to another person; for instance, role playing without an audience or without an awareness of an audience. Both variations are not just early phases which have to be overcome, but important ingredients of striving for optimal empathic understanding.

Fourthly, the affective phase or facet of the mental dimension of empathy (empathic resonance), which is based on emotional and bodily-felt experiences, but which is not proto-empathy anymore, may be expressed verbally as well as non-verbally. The expressed embodiment of affective empathy may be received and responded to in body language.

The next remark concerns the typology as such. In this text the focus is on conceptualizing empathic understanding. However, in real life empathic understanding is a process. The facets of the mental and social dimensions and the combinations of these can also be seen as the phases of a circular or rather spiral process of understanding each other.

According to our discussions above this typology has to be understood in the following way. The affective type of empathic understanding is basic and influences the cognitive and (hermeneutic) interpretive type. The interpretive type of empathic understanding is nourished by the affective and cognitive types. Thus, besides the cognitive component the interpretive type has an affective component. This affective component may be expressed or not. It may be expressed verbally or non-verbally. The embodiment of affective empathy may also be not expressed or expressed, noticed (received) or interactively exchanged. In addition, the expressed type of empathic understanding influences the received and interactive types. The received type influences the interactive type. Hence, the interactive type of empathic understanding is nourished by expressing and receiving empathy. (Of course, there are mutual influences between two mental or two social types of empathy. These and other reciprocal relationships between types are not expressed in Figure 1 for reasons of simplicity of presentation.) Therefore, a combination of the interpretive and the interactive types of empathic understanding will deliver the most optimal variation of empathic understanding. We call this type or variation dialogical-hermeneutical empathic understanding. This type of understanding implies self-clarification on the part of the interpreting empathizer because his or her for-understandings or pre-suppositions are being confronted by the behavior of the researched; the dialogical character implies that the interactions show mutual respect and appreciation, a double hermeneutics, openness and a striving for communicative symmetry, etc.[56,57] The dialogical-hermeneutic type of empathic

124

understanding also includes a narrative dimension of optimal empathic understanding. Josselson, for instance, says:

> "The empathic stance orients us as researchers to other people's experience and meaning-making, which is communicated to us through narrative. To understand another within the empathic stance means being able to understand their stories."
> (Ref. 58, p. 32)

Thus, the conceptual structure in Figure 1 expresses that optimized empathic understanding always has both mental facets (or phases) and social facets (or phases). This conceptualization of optimal empathic understanding may also help to evaluate the quality of empathic understanding. In this sense this 3x4 typology may be seen as a taxonomy. Conceived in this way empathic understanding does have an intrinsical device for being checked, tested, reformulated and developed further. This built-in self-validation implies a two-way device: a mental way, the hermeneutical circle, and a social way, the interactive process. These two self-validating processes should be intertwined. Within this intertwining process of validation cognitive, interpretive as well as affective moments play their part.

Empathic understanding as conceptualized above has a special significance for human research when this research has a participatory character, an emancipatory aim or a goal of empowerment. Indeed, the social dimension implies that the other, the empathizee, may develop his or her self-understanding, may develop a better understanding of his or her situation, and may have the experience of being taken seriously and being respected.[56,59]

5. Imagination

We have defined empathic understanding as placing oneself *imaginatively* in another's experiential world while feeling into her or his experiences with the aim of comprehending these experiences. The question may arise how imagination is understood in this context and how the quality of imagination contributes to the quality of empathic understanding. Can the imagination be developed to reach an optimal quality that facilitates dialogical-hermeneutical empathic understanding?

According to the philosopher Maxine Greene "imagination is what, above all, makes empathy possible" (Ref. 60, p. 3), because imagination is the cognitive capacity that permits us to give credence to alternative possibilities. It allows us to break with what we take for granted and to enter into the alien world of another person, to discover how it looks and feels from the vantage point of that other. We need not approve or appreciate it, but we grasp it as a human possibility. We learn to look at things "as if they could be otherwise" (*ibid.*, p. 19); we break with what is supposedly fixed and finished from our personal point of view. Greene's conception of the imagination comes very close to that of the philosophers John Dewey and Iris Murdoch, who hold that imagination results from the consciousness of a gap between the actual and the possible, from the awareness that our reality goes beyond 'mere fact'. According to Murdoch we can only reach reality through the strength and refinement of the imagination, which she describes as "a type of reflection on people, events, etc., which builds detail, adds colour, conjures up possibilities in ways which go beyond what could be said to be strictly factual." (Ref. 61, p. 198) From these perspectives, we come to the definition of imagination as *an exploration of possibilities that transcend the boundaries of 'mere fact'.* (cf. Refs. 62–63) How does this view on the imagination contribute to our understanding of empathy and empathic understanding? To answer this question, we will first describe the imagination as a cognitive capacity that is part of the mental dimension of empathic understanding. Secondly, we will argue that the imagination has a social dimension as well, and can be developed in interaction with other people, thus contributing to the quality of empathic understanding.

5.1. *Imagination as Part of the Mental Dimension of Empathic Understanding*

In his book *The work of the imagination* Paul L. Harris studies the development of the imagination, understood as the consideration of alternatives to reality.[64] Contrary to the developmental psychology of Jean Piaget, he holds that the imagination is a cognitive capacity that is gradually strengthened in the course of development, and that is critical for making causal and moral judgements about an actual outcome. Harris argues that pretend play is one of the earliest and most obvious indices of children's imagination. In doesn't emerge in early infancy, but in the second year of life, when the child has already gained a limited but functional understanding of the way that the world works. Pretend play departs from the real world, but children use their causal understanding of the physical and mental world in their make-believe episodes. Instead of a relapse from the 'realistic understanding' of early infancy, pretend play 'offers a way to imagine, explore and talk about possibilities inherent in reality' (p. 8). In their pretend play, children do not distort the real world, but explore possible worlds.

The psychoanalyst Ernest G. Schachtel stresses the importance of 'focal attention' in exploring the world and its possibilities.[65] With this, he means an attentive and conscious mode of experiencing, in which the various senses can be used. Focal attention is "man's capacity to *center* his attention on an object fully, so that he can perceive or understand it from *many* sides, as clearly as possible" (Ref. 65, p. 251). The use of focal attention reaches its peak in childhood, as shown in children's great curiosity and desire for exploration. It is not self-evident, however, that this desire is retained in later life, and people vary considerably in this regard. In the course of time, people's perception and experience get 'conventionalised': they stick to the familiar and the socially accepted, sometimes even to the point of not being able anymore to see unexplored aspects or surprising facets. Schachtel mentions a variety of phenomena that confront and break through our habitual patterns and thus encourage world-openness. Next to works of art and religious images, he mentions jokes, playful activities, bodily sensations, dreams: all these can bring us in a situation in which the taken-for-granted world gives way to the experience of something new and surprising. They help us to look at something from different angles and combine perspectives in an unexpected way. They stimulate our imagination and creativity. For Schachtel, imagination is a source of profound insight and a way for 'widening the scope of human life' (Ref. 65, p. 308).

How does imagination work? Dewey, in *Art as experience*,[66] makes clear that it starts with focal attention. It is important to realise, however, that the individual's paying attention to something, isn't an isolated event. Even in the case of a young child, paying attention takes place against a background of meanings that is socially and culturally constructed. Traditions play an important part in this regard. The working of this background material in an actual experience is an imaginative process. Imagination constitutes the only 'entrance' that enables knowledge of a tradition and other memories to enter the experience in process. In the imaginative process, something new comes into being: "that which is given here and now is extended by meanings and values drawn from what is absent in fact and present only imaginatively." (Ref. 66, p. 272). Csikszentmihalyi and Robinson, discuss this interplay between attention, memory and anticipation on the basis of their empirical research on aesthetic experience.[67] The aesthetic experience comes into being when information given through the work of art interacts with information stored in the memory of the observer. The result of this interaction may be a sudden expansion, new combination or reordering of earlier accumulated information, giving rise to various emotions such as delight, joy or awe. A structural component of the aesthetic experience is the process of discovery, that can be related to the work of art itself, but can also have a broader meaning. Attention focused on a work of art, can result in new insights in the human condition.

Although Dewey holds that in art its potential comes to the fore most completely, he explicitly warns against the conception of imagination as a special capacity that is a privilege of the artist. It is rather a quality that inspires all processes in which we really pay attention to our surroundings. From a psychological point of view imagination is a mental process of attention, memory and anticipation based on an 'ideal experimentation': images and meanings can be endlessly combined and reorganised in the imagination. Dewey points out the dramatic character of imagination, through which, at a certain distance from the immediate exigencies and hardships of life, events can be re-enacted and possibilities can be tested in a 'dramatic rehearsal'.[68] Imagination and the cultural background it is rooted in help us to explore reality in ways that do not disrupt our daily life. We may do this by reading a book, watching a movie, or relating to a work of visual art. Imagination, thus developed, is a hermeneutic process, in which the ideal experimentation has the character of a checking and validating process, and in which the societal, cultural and historical aspects are included. It contributes to the understanding and interpreting process by enriching it with new perspectives that can deepen one's understanding of an event or situation.

5.2. *The Social Dimension of Imagination*

Imagination as a cognitive capacity is not necessarily social in nature: the imagination may be restricted to the world of objects or abstract thoughts. Yet, we have already indicated that there is a close relationship between imagination and empathic understanding. We quoted Greene who holds that imagination makes empathy possible. (This holds for the cognitive and interpretive types. Affective empathy is more basic than imagination.) According to Dewey as well, the imagination enables us to take the perspective of others and to understand their desires and aims, their interests and modes of response. Indeed, the former view of developmental psychology on the children's imagination as a capacity that tempts them toward 'autistic' thinking, is not supported by the outcomes of empirical research.

According to Harris, the capacity to place oneself imaginatively in another's experiential world is related developmentally to role play as an important category of pretend play. Role play is defined by Harris as "pretend play in which the child temporarily acts out the part of someone other than the self using pretend actions and utterances" (Ref. 64, p. 30). The role playing child often acts as if it experiences the world from the perspective of the other person or being (perspectival shift). Role play depends upon a process of simulation: the child projects him- or herself into the make-believe situation faced by a given protagonist. It imagines itself in that same situation and acts vis-à-vis that imaginary situation. The child makes use of its own knowledge and planning mechanisms to fill in the role it adopts.

Children that have a lot of practice in role playing have greater social capabilities than children that don't, in the sense that they are more insightful about mental states. They have learned to imagine the situation that confronts another person and can predict his actions or utterances by formulating a plan or conclusion in relation to that imagined situation, and projecting it onto the person in question. Role play and prediction both call on a simulation process; in the case of role play the child enacts what someone might think or say, in the case of social interaction it may predict so. Children vary in the extent to which they engage in role play, this depends on a variation in opportunities and encouragement, but also on variation in inclination. There seem to be early and stable endogenous differences in the disposition toward simulation and role play. Yet, training is possible. For example, children who have been trained in sociodramatic play are more adroit at mental state understanding.

In their role play, children clearly show that imaginary premises can drive their emotional system, including its physiological components. This is not because of any confusion about what is real and what is imaginary, but it is due to their capacity for absorption in a pretend world — a capacity they share with adults who read a book or watch a movie. In the course of development, however, people learn techniques that increase or decrease involvement. They set in motion some type of override that inhibits the process of identification. One of the reasons that human beings are so emotionally responsive to imagined events may be, that it helps us to understand reports of events we have not witnessed ourselves, thus strengthening our social engagement. We construct in our imagination mental models of scenes and events that are described in narrative reports of others. Our ability to build mental models is primarily directed at the processing of honest testimony with all its emotional and interpersonal ramifications.

Assuming that children use their imagination from an early age to take the perspective of other people, that their capacity to become emotionally involved in a make-believe situation strengthens their social engagement, and that they learn techniques to regulate their involvement, can we think of a development leading from early role play to the most optimal variation of empathic understanding? What would the role of the imagination in the various stages or phases of this development be?

We have already discussed the importance of the imagination for the cognitive and interpretive types of empathy. With regard to the social types of empathy, it is important to distinguish between non-empathic imagination (e.g., pretend play with inanimate objects, not endowed with mental states) and imagination in the 'not expressed' empathic style, the zero-value of the social dimension (e.g. a purely mental conversation with imaginary companions or role play in which no actual other is involved). As the research of Harris shows, imagination in the non-expressed mode still has its importance for developing the really social modes of the imagination. The imagination can be developed in this direction, when the person seeks to articulate the possibilities he or she imagines ('expressed to another'), when the other is responsive to this expression ('received'), and when the imagined possibilities can be explored further in the interaction between them ('interactional'). In a health care situation for example, a therapist may make a perspectival shift thanks to her imagination; she may express her understanding of the experiential world of the other by finding a way for expressing the perspective of the other she has understood imaginatively; depending on the way she expresses her understanding the client may respond to it, and finally, therapist and client may both deepen their understanding of the situation by exploring different perspectives on it in their interaction.

Imagination that is interpretive and interactional contributes to dialogical-hermeneutical empathic understanding in two ways. First and basically, as we explained earlier, it is a prerequisite for placing oneself in another's experiential world by allowing us to break with what we take for granted from our own perspective. Secondly and additionally, it nourishes empathic understanding with perspectives that are not immediately given in the experiential world of the other, but that are derived from the societal, cultural and historical context of the interaction. This may contribute to the potential for change implied in dialogical-hermeneutical empathic understanding, by giving a new impetus to the process of checking, testing and reformulating interpretations and thereby to the change of the participating persons.

6. Worldview

We argued that imagination is a hermeneutical process that contributes to empathic understanding by enriching it with new perspectives that can deepen one's understanding of an event or situation. We mentioned works of art as facilitators of this process. Of course, the relationship of imagination and art is much more complex, for we already need the imagination to construct and appreciate a work of art. We will now argue that worldviews have such a complex relation to imagination as well. We do not mean this in a reductionist sense. We do not aim to clarify the origins of worldviews, or to state that they are nothing but imagination. Still, we hold that worldviews can both be better understood and become more meaningful in a person's life when the role of the imagination is accounted for. How can we grasp the meaning of a worldview?

A worldview (or philosophy of life) can be conceptualized as a reflective-cognitive orientation towards human life. Within this orientation we differentiate between the internal, conceptual or substantial meaning-dimension and the external, functional meaning-dimension. The internal, conceptual meaning-dimension has four aspects: the ontic, the epistemic, the valuative and the existential aspect. The ontic aspect refers to questions and answers concerning what exists or does not exist as well as how something or somebody exists. For instance, humans may be seen as intelligent animals or as hermeneutic beings. This ontic aspect has to do with images of man and pictures of our world or the cosmos. The epistemic aspect refers to what can be known, understood, researched or explained. For instance, can we only know what is rational and observable, or can we know more than this? The valuative (or normative in a broad sense) aspect refers to moral, ethical, esthetical and other evaluative questions and answers. For instance, human behavior may be judged as ethically wrong, an article of dress may be judged as beautiful and the argumentation of a speaker may be evaluated as very strong. The existential aspect refers to questions and answers concerning the meaning of life as a whole. Where do we come from? What is the ultimate goal of my life? This existential aspect includes spiritual and religious variations of it. A worldview, a reflective-cognitive orientation towards human life as indicated above, may have a certain degree of internal integration or coherence. This degree of coherence influences the way in which a worldview may function in everyday life.

The external, functional meaning-dimension of a worldview has to do with the value, utility or yield of a worldview for a single person or a group.

In psychology and sociology, (religious) worldviews are often seen as important sources of meaning, because they function as a frame of reference to understand life events that can't be explained in another way. As interpretive frames of reference worldviews enable people to see order in what happens to them and to experience a sense of control over their lives.[69,70] Order and control are no doubt important aspects of finding meaning in life. Of course, if a worldview has little coherence its function of delivering order or control will be weak. In addition, we think that we underestimate the power of worldviews, when we only conceive of them as an explanatory or ordering frame. When people focus on explaining and controlling their world, they risk to lose their openness to what is strange and unfamiliar. Worldviews, whether they are religious or not, run the risk of becoming fixed and dogmatic. This may be enforced by an institutionalization of a worldview, which may be accompanied by a 'worldview regime' (by analogy with a political or a religious regime). Such a regime not only has a protective role, but also a suppressive function. If that happens, empathic understanding will be blocked, at least empathic understanding of strange, unfamiliar or foreign people. On the other hand, we need not reduce a worldview to a set of answers to existential and moral questions nor to a suppressing institution. Indeed, it is not necessarily the case that a worldview only satisfies the need for order, safety, security or control and can not satisfy the need for freedom, development, exploration or imagination. A worldview is also a source of existential questions that may evoke wonder and awe in us. It can open us to the richness of the world or can make us despair about injustice and cruelty. It can make us care for the fate and experiential world of others. In addition, the meanings of which a worldview may be the source concern ontic, epistemic, valuative and existential aspects of a worldview. If a worldview gives room to the imagination it can help to see or experience more possibilities associated with each of these aspects.

Worldviews may show alternatives to reality to us. Religious, socialist or humanist worldviews may criticize the way we accept or even contribute to injustice, oppression, violence and poverty in the world we live in. They may offer perspectives of a better world, on both an individual and a global level. They may inspire people to change their lives and to commit themselves to changes in their society. As we have seen, the consideration of alternatives to reality is the quintessence of the imagination. Worldviews stimulate people's imagination. In this way, a worldview is a source of meaning in life, not by functioning as an explanatory frame, but by offering possibilities one may explore or devote oneself to.

Clarifying and understanding one's own worldview enables one to recognize and gain insight into the worldview aspects in the narrative of another. It is a hermeneutical process in which one needs understanding of one's own values, beliefs and orientation in life to be able to understand those of others. Both reflectivity and imagination are necessary for this hermeneutical process, in which the person doesn't try to convince the other, but tries to empathically understand him or her. In this dialogical process, the understanding of both one's own worldview and that of the other will be deepened. For this to happen, it is important that the person doesn't cling to a fixed frame of reference, but is open to new perspectives that may broaden or deepen his or her outlook on life. Entering imaginatively into a book, a film, a poem etc. can help to acquire insight and openness with regard to one's worldview. This gives evidence of the complex relationship between imagination and worldview: the imagination contributes to an open worldview that stimulates one's imagination.

Under what conditions can the imaginative power of worldviews be released? In the first place, a person must be able to relate her own experiences to the insights a worldview offers. Only in this way can it become personally meaningful and contribute to self understanding. A worldview may offer a perspective to reflect on one's experiences and to empathically relate to oneself as another (self empathy). In the second place, a worldview must contribute to a trusting relationship between the person and his world. Only on the basis of trust can one open oneself sufficiently to the world to acknowledge strange perspectives and empathically relate to other people. In the third place a worldview must inspire people to imaginatively explore new possibilities, by which a broader view on reality can be opened. Symbols, rituals, art and narratives are of great importance in this regard. Finally, the worldview must offer possibilities for critically testing one's personal experiences, values and goals in life. People can be mistaken in what they strive for and commit themselves to. They can attach value to something that doesn't deserve it on the basis of a personal desire, vulnerability or anxiety. According to the philosopher Charles Taylor,[71] the only way to check this is in the confrontation with criticism, in a dialogue with one's social surroundings and the cultural tradition one belongs to. Commitments are developed in a complex social structure, and can't be changed by reason alone. Yet, difficult though it may be, the human capacity to reflect on emotions and preferences and to evaluate these enables us to take responsibility for our way of dealing with them. The capacity to relate empathically to the emotions, feelings and thoughts of others, enables us to be compassionate. Responsibility and compassion express themselves in care for our surroundings. This is a learning process that can be facilitated in dialogue with one's worldview, which offers ways to articulate and check one's central values.

Worldviews do not contribute to dialogical-hermeneutical empathic understanding as a matter of course. Apart from the conditions just mentioned, the person needs to choose for a reflective, critical and open attitude towards one's own worldview. This is an act of the will, the will to believe and take responsibility on the basis of that belief, *and* the will to be self-reflective and self-critical.

Worldviews aren't given once and for all, especially not in our late modern society. They must be developed and taken care of, or they will either fossilize or crush to a subjectivism in which only individual preferences count. Empathic understanding and imagination flourish neither in a context of rigidity nor in a context of subjectivism. A worldview that contributes to the articulation and evaluation of one's own central values, protects a person from 'sham-empathy' in which she uncritically identifies herself with the values of another. A dialogue between different worldviews may lead to a hermeneutical process, to empathic understanding in its richest form. It may contribute to an exchange and exploration of perspectives and possibilities that facilitates individual and social change. Such a dialogue can both be a source of self-understanding and personal meaning in life as of a profound understanding of the other and his or her meaning in life. Such a dialogue is needed for (scientific) understanding, policy-making and meaningfully participating in a multiform society.

7. Conclusion

In this article a comprehensive conceptualization of empathic understanding is developed which makes empathy more worthwhile for doing human research as well as acting in other professional contexts. To do so, we first discussed the most important criticisms on empathic understanding (Fay, empirical-analytical philosophers of science, some hermeneutical philosophers like Gadamer and Ricoeur). We came to the conclusion that these critics conceive empathy very narrowly or one-sidedly, while also neglecting or underestimating the importance of the context of discovery, the methodological domain, the practice of doing empirical research as well as everyday practices and professional practices such as health care, education and counseling. To come to a conceptualization that includes an idea of opti-

134

mal empathic understanding, we turned our attention to appreciations of empathic understanding (Schleiermacher, Dilthey, Weber, Schutz, Winch, phenomenological philosophers like Husserl, Stein, Merleau-Ponty, Duyndam, psychotherapists and counselors like Rogers, Kohut and Gendlin, and methodologists of social scientific research, like Gordon). We conclude from these positive views that empathy is indispensable for understanding others in ordinary life, for keeping personal mental health and for maintaining human societies. We define empathic understanding as *placing oneself imaginatively in another's experiential world while feeling into her or his experiences with the aim of comprehending these experiences.* Optimizing the development of empathic understanding needs communication. Therefore, a conceptualization of optimal empathic understanding must include a socially behavioral dimension in addition to the dimension of attitudinal or mental ability. In the mental dimension, three types of empathic understanding can be distinguished: the affective type, the cognitive type and the interpretive type. In the social dimension, four types of empathic understanding are distinguished: not-expressed to another, expressed to another, received and interactional. We have constructed a cross product of these sets or dimensions (see Figure 1), that represents a typology of twelve types of empathic understanding of which the dialogical-hermeneutical type is the most optimal one.

As we have defined empathic understanding as placing oneself *imaginatively* in another's experiential world, our conceptualization needs an optimizing explication of the term 'imaginatively'. We define imagination as *an exploration of possibilities that transcend the boundaries of 'mere fact'.* We came to the conclusion that imagination contributes to dialogical-hermeneutical empathic understanding in two ways. As part of the social dimension of empathic understanding, it is a prerequisite for placing oneself in another's experiential world by allowing the empathizer to break with what he takes for granted from his own perspective, and express his understanding in a way the other can respond to. As part of the mental dimension, it nourishes empathic understanding with perspectives that are not immediately given in the experiential world of the other, but that are derived from the societal, cultural and historical context of the interaction. Worldviews are part of this context and can contribute to empathic understanding. A worldview, understood as a reflective-cognitive orientation towards human life, may show alternatives to reality to a person and stimulate her imagination. We specified some conditions under which the imaginative power of worldviews can be released. Furthermore, a dialogue between different worldviews may lead to a better understanding of both one's own values, beliefs and orientation in life, and those of others. This is of the utmost importance in the complex world we live in. Multiculturalism, migration, globalization, the rapidity of social change, fragmentation, ICT, etc., make it necessary that not only professionals must develop and train their empathic competence. We think that, given the complexity of contemporary societies, empathic understanding should be explicitly developed and trained by everyone. As for us empathic competence should belong to the objectives of basic educational programs, all over the world.

References

1. B. Fay, *Contemporary philosophy of social science.* Oxford, UK: Blackwell (1996).
2. K. R. Popper, *Objective knowledge.* Oxford: At the Clarendon Press (1972).
3. C. G. Hempel, *Aspects of scientific explanation and other essays in the philosophy of science.* New York: The Free Press (1965).
4. E. Nagel, *The structure of science. Problems in the logic of scientific explanation.* New York: Harcourt, Brace & World (1961).
5. R. S. Rudner, *Philosophy of social science.* Englewood Cliffs, N.J.: Prentice-Hall (1966).
6. A. Ryan, *The philosophy of the social sciences.* London: Macmillan (1970).
7. Th. Nickles (Ed.), *Scientific discovery, logic and rationality.* Dordrecht/Boston: Reidel (1980).
8. H.-G. Gadamer, *Truth and method.* London: Sheed and Ward (Originally in German: *Wahrheit und Methode*, Tübingen, 1960) (1975).
9. H.-G. Gadamer, On the circle of understanding. In: J. M. Connolly and T. Keutner (Eds.), *Hermeneutics versus Science?* Notre Dame, Indiana: University of Notre Dame Press, 68–78 (1988).
10. P. Ricoeur, *Hermeneutics and the human sciences.* Cambridge: Cambridge University Press (1981).
11. M. Alvesson, Beyond neopositivists, romantics, and localists: a reflexive approach to interviews in organizational research. *Academy of Management Review* **28** (1), 13–33 (2003).
12. C. Seale, Qualitative interviewing. In: C. Seale (Ed.), *Researching society and culture.* London: Sage, 202–216 (1998).
13. D. Silverman, The impossible dreams of reformism and romanticism. In: J.F. Gubrium and D. Silverman (Eds.), *The politics of field research. Sociology beyond enlightenment.* London: Sage, 30–48 (1989).
14. Fr. D. E. Schleiermacher, *Hermeneutics.* (Ed. and with an introduction by Heinz Kimmerle.) Heidelberg, Carl Winter: Universitätsverlag (1959).
15. R. E. Palmer, *Hermeneutics.* Evanston: Northwestern University Press (1969).
16. W. Dilthey, *Collected papers.* Göttingen: Vandenhoeck & Ruprecht (1913–1974).

17. M. Weber, *The methodology of the social sciences*. New York: Free Press (1949).

18. M. Weber, The interpretive understanding of social action. In: M. Brodbeck (Ed.), *Readings in the philosophy of the social sciences*. New York: MacMillan, 20–33 (1968).

19. A. Schutz, *Collected papers I* (Ed.: M. Natanson). The Hague: Martinus Nijhoff (1962).

20. A. Schutz, *Collected papers II* (Ed.: A. Brodersen). The Hague: Martinus Nijhoff (1964).

21. P. Winch, *The idea of a social science and its relation to philosophy*. London: Routledge & Kegan Paul (1958).

22. E. Stein, *On the problem of empathy*. Washington, DC: ICS Publications (Kluwer) (Third revised edition; first English edition in 1962; first German edition in 1917) (1989).

23. E. Husserl, *Ideas: general introduction to pure phenomenology*. New York: The Macmillan Company (First English edition in 1931; first German edition in 1913) (1952).

24. E. Husserl, *Cartesianische Mediationen und Pariser Vorträge*. The Hague: Martinus Nijhoff (1950).

25. M. Merleau-Ponty, *Phenomenology of Perception*. London: Routledge & Kegan Paul (First French edition in 1945) (1962).

26. M. Merleau-Ponty, *The Visible and the Invisible*. Evanston: Northwestern University Press (First French edition in 1964) (1968).

27. J. Duyndam, Empowerment by empathy. On good and gruesome empathy. In: A. Halsema and D. van Houten (Eds.), *Empowering humanity. State of the art in humanistics*. Utrecht: De Tijdstroom, 139–148 (2002).

28. C. R. Rogers, *On becoming a person. A therapist's view of psychotherapy*. Boston: Houghton Mifflin (1961).

29. C. R. Rogers, The interpersonal relationship: the core of guidance. In: C. R. Rogers and B. Stevens (Eds.), *Person to person. The problem of being human*. New York: Pocket Books, 85–101 (1967).

30. C. R. Rogers, *Client-centred Therapy. Its current practice, implications and theory*. Boston: Houghton Mifflin (1951).

31. G. A. M. Widdershoven, Hermeneutics and client-centred psychotherapy. In: A. W. M. Mooij and G. A. M. Widdershoven (Eds.), *Hermeneutics and psychology*. Meppel/Amsterdam: Boom, 74–98 (1992).

32. J. D. Bozarth, Empathy from the framework of client-centered theory and the Rogerian hypothesis. In: A. C. Bohart and L. S. Greenberg (Eds.), *Empathy reconsidered*. Washington, DC: APA, 81–102 (1997).

33. E. T. Gendlin, The experiential response. In: E. Hammer (Ed.), *The use of interpretation in treatment*. New York: Grune & Stratton, 208–227 (1968).

34. Ch. H. Cooley, The roots of social knowledge. *The American Journal of Sociology* **32**, 59–79 (1926).

35. D. Bru, Empathy: more than an attitude. *Journal for Humanistics* **2** (5), 50–57 (2001).

36. H. Kohut, *How does analysis cure?* Chicago: The University of Chicago Press (1984).

37. H. Kohut, Introspection, empathy, and psychoanalysis: an examination of the relationship between mode of observation and theory. In: P. H. Ornstein (Ed.), *The search for the self.* New York: International Universities Press, Vol. 1, 205–232 (1959).

38. H. Kohut, On empathy. In: P. H. Ornstein (Ed.), *The search for the self.* New York: International Universities Press, Vol. 4, 525–535 (1981).

39. D. S. MacIsaac, Empathy: Heinz Kohut's contribution. In: A. C. Bohart and L. S. Greenberg (Eds.), *Empathy reconsidered.* Washington, DC: APA, 245–264 (1997).

40. R. van den Hoofdakker, Empathy. *Journal for Humanistics* **2** (5), 7–15 (2001).

41. W. van Tilburg, Conditions for empathy. *Journal for Humanistics* **2** (5), 24–31 (2001).

42. A. C. Bohart and L. S. Greenberg (Eds.), *Empathy reconsidered. New directions in psychotherapy.* Washington, DC: American Psychological Association (1997).

43. H. J. Rubin and I. S. Rubin, *Qualitative interviewing. The art of hearing data.* Thousand Oaks: Sage (1995).

44. S. Kvale, InterViews. *An introduction to qualitative research interviewing.* Thousand Oaks: Sage (1996).

45. R. L. Gordon, *Interviewing: Strategies, Techniques and Tactics. Homewood.* Ill.: The Dorsey Press (1980).

46. N. Eisenberg and J. Strayer, Critical issues in the study of empathy. In: N. Eisenberg and J. Strayer (Eds.), *Empathy and its development.* New York: Cambridge University Press, 3–13 (1987).

47. V. Gallese and A. Goldman, Mirror neurons and the simulation theory of mind-reading. *Trends in Cognitive Science* **2**, 493–501 (1998).

48. V. S. Ramachandran, Mirror neurons. `www.edge.org/documents/archive/edge69.html` (2000).

49. G. Rizzolatti, L. Fadiga, L. Fogassi and V. Gallese, From mirror neurons to imitation: facts and speculations. In: W. Prinz and A. Meltzoff (Eds.), *The imitative mind: development, evolution and brain bases.* Cambridge: University Press, 247–266 (2002).

50. A. Barnes and P. Thagard, Empathy and analogy. `http://cogsci.uwaterloo.ca/Articles/Pages/Empathy.html` (1997).

51. H. H. Kögler and K. R. Stueber (Eds.), *Empathy and agency. The problem of understanding in the human sciences.* Boulder, Colorado, US; Oxford, UK: Westview Press (2000).

52. A. Smaling, The meaning of empathic understanding in human inquiry. Paper presented at the 55th International Phenomenology Congress, Nijmegen, The Netherlands (2005, August).

53. P. M. van Strien, *Empathy.* Amsterdam: Thela-Thesis (1999).

54. J. C. Tronto, *Moral boundaries: a political argument for an ethic of care.* New York: Routledge (1993).

55. A. Giddens, *New rules of sociological method.* London: Hutchinson (1976).

56. A. Smaling, Qualitative interviewing: contextualization and empowerment. In: I. Maso and F. Wester (Eds.), *The deliberate dialogue.* Brussels: VUB-press, 15–28 (1996).

57. A. Smaling, Dialogical partnership — The relationship between the researcher and the researched in action research. In: B. Boog, H. Coenen, L. Keune and R. Lammerts (Eds.), *The complexity of relationships in action research.* Tilburg: Tilburg University Press, 1–15 (1998).

58. R. Josselson, Imagining the Real. Empathy, Narrative, and the Dialogic Self. In: R. Josselson and A. Lieblich (Eds.), *Interpreting Experience: the narrative study of lives.* London: Sage, 27–42 (1995).

59. A. Smaling and I. Maso, The humanist potentialities of qualitative research: Inner growth, dialogical relationships, empowerment and normative professionalism. In: A. Halsema and D. van Houten (Eds.), *Empowering humanity. State of the art in humanistics.* Utrecht: De Tijdstroom, 37–60 (2002).

60. M. Greene, *Releasing the imagination: Essays on education, the arts, and social change.* San Francisco, CA: Jossey-Bass (1995).

61. I. Murdoch, Existentialists and mystics: Writings on philosophy and literature. New York: Allen Lane The Penguin Press (1997).

62. H. A. Alma, Ground breaking exploration: an exploration of imagination from the perspective of psychology of religion. *Nederlands Theologisch Tijdschrift* **56**, 115–129 (2002).

63. R. Kearney, *The wake of imagination: Ideas of creativity in Western culture.* London: Hutchinson (1988).

64. P. L. Harris, *The work of the imagination.* Oxford: Blackwell (2000).

65. E. G. Schachtel, *Metamorphosis: On the development of affect, perception, attention and memory (with a preface by J. L. Singer).* New York: Da Capo Press (Original edition 1959) (1984).

66. J. Dewey, *Art as experience.* New York: Perigee Books (Original edition 1934) (1980).

67. M. Csikszentmihalyi and R. E. Robinson, *The art of seeing: An interpretation of the aesthetic encounter.* Los Angeles: The J. Paul Getty Museum (1990).

68. J. Dewey, *Experience and nature* (The later works 1925–1953: Vol. 1, 1925). Carbondale: Southern Illinois University Press (1981).

69. R. F. Baumeister, *Meanings of life.* New York: The Guilford Press (1991).

70. B. Spilka, P. Shaver and L. A. Kirkpatrick, A general attribution theory for the psychology of religion. *Journal for the Scientific Study of Religion* **24**, 1–20 (1985).

71. C. Taylor, *Sources of the self: The making of the modern identity.* Cambridge: Cambridge University Press (1989).

WORLDVIEW AS RELATIONAL NOTION? RECONSIDERING THE RELATIONS BETWEEN WORLDVIEWS, SCIENCE AND US FROM A RADICAL SYMMETRICAL ANTHROPOLOGY

LIEVE ORYE

Centre for Intercultural Communication and Interaction
University of Gent, Belgium
E-mail: lieve.orye@UGent.be

I develop an anthropological reflection on the notion of worldview. Pointing out the connection often made between worldview and map, orientation and globality, as well as the frequent discussions in (Christian) philosophical reflection on the worldview notion of issues of relativism and a swinging between conscious commitment and unconscious bias, the possibility is discussed that the notion used in such reflections is problematic. On the basis of Ed Hutchins work on ship navigation and the reflections of two reviewers of his work, Bruno Latour and Tim Ingold, a foundational error in human sciences is pointed out and the possibility of considering the notion of worldview as a relational notion is discussed.

1. Worldview, Map and Orientation

The notion of worldview has a recent history. The term usually invokes a reflection on what one's own worldview might be, which elements one would see part of it, and which elements one would reject. Other reflections are about how different one's own worldview seems to be from others and how a dialogue with people living their life from within a different worldview might be possible. Reflection on the notion of worldview itself is rather rare and usually takes the form of philosophy. Very recently, the notion has been more extensively scrutinized by two Christian thinkers, David Naugle and James Sire. Naugle is professor of philosophy at Dallas Baptist University, Sire is campus lecturer for Intervarsity Christian Fellowship, an evangelical campus mission for students. His book on the concept of worldview is written from this position, builds upon Naugle's study of the history of the concept and finally results in apologetics. This latter apologetic argument does not concern us here. Sire's *Naming the Elephant* will be taken as material for a more anthropological reflection on the notion.[1] I will draw on radical symmetrical anthropological work to argue for a clear shift in our thinking about worldview, science and us, a shift that results in the challenge to think worldview no longer in a relation-blind and inside-the-mind-versus-the-world way.

The notion of worldview is often connected with such notions as orientation and map. Apostel, whose views on worldview are outlined in the first volume of this series,[2] sees worldview as a map that people use to orient and explain, and from which they evaluate and act, and put forward prognoses and visions of the future. This statement in itself does not allow one to conclude whether he viewed this use of worldview as a rather specific activity done only by some persons placed in a position or situation to do so, or whether having a worldview to orient oneself is considered a human universal, a description of a universal human condition. Quite often the latter is assumed. As Sire frequently notes, a worldview is simply inescapable:

> "[I]n an important sense [...] everyone [...] has a worldview. Any rejection of that notion is self-refuting. It's like saying, 'There are no absolutes; everything is relative,' a statement that, if true, is false, in other words, self-referentially incoherent." (Ref. 1, p. 29)

His and Naugle's work show how a discussion of this notion is often linked in the same breath with issues of relativism and a major part of their argument is that a Christian worldview somehow escapes that trap. Knowledge is perspectival, there is no view from nowhere, there is no neutral ground. Postmodernism is congratulated by these thinkers for showing this and the step towards a relativization of science as just one more worldview based on a leap of faith is easily made. The last step, however, of denying objective reality with its own integrity and attendant truth, is refused and postmodernist thinkers end up vilified after all. But what one often finds in such discussions is a going back and forward between worldview as something constructed, held consciously and involving a commitment and worldview as an unconscious or built-in bias or basic cognitive framework. I will discuss the possibility that such swinging is symptomatic for a relation-blind notion of worldview and will show how taking its relational character seriously might open up quite a different picture.

The relational character of this notion will be brought to the foreground by discussing the notion of map and cartography in anthropological work. A description of worldview as map usually does not clearly define what kind of map one has in mind. Is it similar to the ones cartographers provide us with or does it more closely resemble such maps that rather narrate through drawings how to go from here to there? The common denominator is that all maps contain translations of elements in the world into a different medium, such as paper or perhaps a mental space. However, the relations between these elements and the way of navigation or orienting oneself can take different forms. Anthropology provides us here with interesting data that warns against easy generalization. As the authors of *Cartographic Encounters* point out we should be aware that we easily bring in our Eurocentric assumptions about mapping which are connected to strict distinctions between nature and culture, image and text, terrain and cosmology, time and space, etc.[3] As a consequence, we have limited our ability to understand and even recognize non-Western 'maps'. Even more, with anthropologists Ed Hutchins, Bruno Latour and Tim Ingold one can point out that we have even limited our ability to understand modern cartography and navigation. If map then is an important metaphor to understand the concept of worldview, it is quite possible that the same assumptions are present. The fact that worldview as orientation and map is usually seen to have a characteristic of globality might be a case in point. In a Think Piece in *Zygon*, a journal on religion and science, Gregory Peterson defines religion as orienting worldviews that are fundamental, explanatory and global in character. Religion involves conceptions of a general order of existence. It tells us about the world and how it works by constructing a system of symbols that inform an individual's behavior, goals and aspirations. Peterson recognizes that globality is not a universal feature of religion, but notes that it certainly is characteristic of the historically successful (i.e., widespread) religious traditions. In *Naming the Elephant* Sire illustrates this character of globality by starting his book with the following story. A little boy comes home and tells his father that the teacher had shown them a big round globe. It was said to be the model of the world, a world surrounded by space. Puzzled, the boy then asks his father what holds up the world? 'A camel' is the father's answer but both are soon caught up in a *regressus ad infinitum*, 'it's elephant all the way down.' Sire's book is build around this story. However, he does not note the specificity of the school situation in which children get to see a model of the world as a globe. He does note that the boy's question and the father's answer are informed by a worldview deeply involved with the society in which the boy and the father live, but the globe as model for the world is simply taken for granted (Ref. 1, p. 79). Nevertheless, such a situation is quite specific. As Ingold notes about the schoolchild looking at the globe, "with the world as globe, far from coming into being in and through a life process, it figures as an entity that is, as it were, presented to or confronted by life" (Ref. 4, p. 32).

In what follows I want to argue that the concept of worldview as it figures in Sire's work is colored by a fundamental problem. It draws together what might have to be considered apart and generalizes too easily from what is a quite specific situation and relation. Recognizing worldview as a relational notion will allow us to avoid this mistake and to open up a study of worldview practices as part of anthropology as 'a science of engagement in a relational world'.

2. Maps and Particular Ways of Relating

2.1. *A Foundational Error and Irony*

In 1995 Edwin Hutchins described and discussed the way a navy ship could orient itself and find its way into the harbor. The study launched the notion of distributed cognition and broke away from the narrow views on cognition as something between the ears, an 'in the head only' affair. Focusing his attention on the navigation team, the latter was studied as a cognitive and computational system in its own right. A rich reality becomes visible in which charts, instruments and people are formed and connected in a dense web that allows the ship to harbor safely. The study culminates in a strong critique of existing cognitive anthropology:

> "In the quest to learn what people know, anthropologists lost track both of how people go about knowing what they know and of the contribution of the environments in which the knowing is accomplished." (Ref. 5, xii)

In his last chapter Hutchins criticizes both cognitivists for marginalizing culture and anthropologists for agreeing with and even enhancing such a marginalization. Cognitivists started from the premise that one could first attempt to understand the individual and than bring culture into the picture. For Hutchins (Ref. 5, p. 354), however, "[t]he understanding of the individual that has developed without consideration of cultural process is fundamentally flawed." However, rather than simply proposing an alternative that adds culture, context and history, Hutchins concludes that "some of what has been done in cognitive science must now be undone so that these things can be brought into the cognitive picture." A mistake was made and needs to be rectified before we can go on.

"If there are fundamental deficiencies in the dominant conceptions of cognition in cognitive science, how did that come about?" (Ref. 5, p. 356). Hutchins is aware that his work isn't easily digested by many scholars of cognition. "[S]ome of the unexamined assumptions of the field make my words unruly," he claims, and to give his new framework a chance, he deconstructs the old one through contrasting the 'Official History of Cognitive Science' with an alternative one. According to the official story, the computer was made in the image of the human. His alternative story however narrates how the model of cognition as symbol manipulation is a model based on the cognitive properties of a socio-cultural system in which the human actor has been removed, only to be taken afterwards as a model of individual cognition projecting much of the socio-cultural system inside the individual's head. That is, charts and maps easily became mental maps and models. However, there is something highly ironic about the view on human beings that builds upon such projections. "The computer was made in the image of the sociocultural system, and the human was remade in the image of the computer" (Ref. 5, p. 364). That is, at first, a person, more specifically a mathematician or logician, who was actually doing the manipulation of symbols with his or her hands and eyes was the model cognitive system (in the search for Artificial Intelligence). An interaction between person and symbols through manual manipulation was considered to do something computational. In such a situation Hutchins notes that the symbols are in the environment of the human and the human is manipulating the symbols. More importantly, the cognitive properties of the human then are not the same as the properties of the system that is made up of the human in interaction with these symbols. He continues by pointing out that Turing's central discovery was that "the embodied actions of the mathematician and the world in which the mathematician acted could be idealized and abstracted in such a way that the mathematician could be eliminated" (Ref. 5, p. 361). That is, the way the mathematician actually interacted with the material world was of no importance to the production of the computation, was "no more than an implementational detail" and could be considered inessential. What was considered essential to the abstract manipulation of symbols was exactly not what was done by the person. Automation of symbol manipulation became a possibility. The cognitive processes and activity of the person as well as all the problems when interacting with material symbols could be factored out. In itself, the automation of symbol manipulation, through a computer, contributes to extending the boundaries of human computational accomplishments and is to be applauded. However, considering such a computer subsequently as a model for human cognition is highly problematic. The human actor and his/her relation to his/her environment is fatally distorted.

> "Having failed to notice that the central metaphor of the physical-symbol-system hypothesis captured the properties of a sociocultural system rather than those of an individual mind, AI and information-processing psychology proposed some radical conceptual surgery for the modeled human. The brain was removed and replaced with a computer. The surgery was a success. However, there was an apparently unintended side effect: the hands, the eyes, the ears, the nose, the mouth, and the emotions all fell away when the brain was replaced by a computer." (Ref. 5, p. 363)

2.2. *One Side of the Coin: from Minds and Forms to Goals and Movement*

Both Bruno Latour and Tim Ingold applauded Hutchins' work and critique of the cognitive sciences. Both, however, had reservations as well. Their concerns are complementary: Latour's concern is that Hutchins had more or less ignored science studies and had made the mistake of overlooking the specificity of navy navigation; Ingold's problem was that Hutchins too quickly generalized from what was to be considered a very specific situation, misrepresenting human reality in general. Where Hutchins argues to replace an 'in the head' approach with an approach that placed the individual firmly in its environment, both Latour and Ingold demand attention for the specific kind of individual–environment relations that can be found in ship navigation. For both, ship navigation is an example of individual–environment relations in which the person as person with a personal history of enskilment is erased. Cognition as computation is a very specific phenomenon.

For Bruno Latour *Cognition in the Wild* is a milestone. Not only does it point out the totally unexpected effects the computer has had on the definition of humans and their capacities, it is also a work that could lead to a complete redefinition of what it is to think. Hutchins takes cognitive science seriously, states Latour, and shows that the computer is a good description of socio-cultural systems, not of human cognitive functions (Ref. 6, p. 54). The book allowed Latour, less than 10 years after his last plea for a moratorium on cognitive explanations of science in *Science in Action* to lift the ban,[7] since "in the meantime, cognitive explanations would have been dissolved beyond recognition by [...] Hutchins and made thoroughly compatible with the social explanations of science, technology and formalism devised by my colleagues and myself" (Ref. 6, p. 62). Hutchins' work chimes in nicely with Latour's own project of developing an anthropology of the sciences. Like Hutchins, Latour battles against assumptions that would make his project inconceivable. For long, science was considered non-anthropologizable and figured as unquestioned standard and point of comparison in anthropology. Developing an anthropology of the sciences meant overthrowing this framework, rethinking such asymmetrical anthropology and human science more broadly and studying science-in-the-making as quite a specific, costly and energy-consuming practice. The latter allowed Latour to throw out a vocabulary in terms of minds and forms of thought and to replace it with a focus on movements and goals in order to explain differences between science and non-science, between cartography and beer card maps. Asymmetry was not denied but seen as to be explained simply by pointing out the displacements and inscriptions involved in scientific practice. The difference between the enunciations of a researcher and for instance a sorcerer, Latour stated, is not a matter of rationality versus irrationality, but lies in the displacement of the researcher and is thus a matter of local action versus the action of displacement and of placing at a distance.[8] This displacement, from a local fieldwork situation to a university setting for instance, necessitates the enunciations to take a rather specific form. Talking to someone, such as one's university colleagues, about things and people who are not immediately around involves specific strategies and instruments that allow enunciations that serve as intermediaries in exchanges, to undergo transformations that no longer need local action. To clarify this, Latour discusses the situation of La Pérouse, a French explorer sent out by Louis XVI with the specific mission to improve the maps of the Pacific. This purpose informs many of the activities on board of the ship and informs the definition of success: what counts is what can be represented on a map using an exact projection in longitude and latitude, and furthermore, success depends as well on the return of that map to France where it can be multiplied through printing. Once the map returns, others can take it and sail to the Pacific, knowing the situation in advance in such a way that faster and safer routes can be taken. Local people, of course, know their own environment, know the form of islands and rivers. The difference, Latour points out, lies in the form of this form. Whereas one of the locals encountered by the crew draws island and river in the sand as a temporary intermediary to communicate the knowledges he has embodied since childhood to the stranger, such medium is useless for La Pérouse who has no intention to stay or embody such knowledge himself. He needs to go back to France with information that can be multiplied, distributed and used by others apart from him. This goal changes the role and form of the intermediaries fundamentally. Whereas in the local situation of people living their life there, intermediaries are only rarely used and are of a temporary form, these become the central focus and involve much energy, time and costs — to give them much more permanence while nevertheless making them easily transportable — as soon as one is collecting and transporting information for people who are not around. A cartographer's map is highly interesting as an example of the latter case. It allows others who have never been in the Pacific to have knowledge beforehand and, even more, it allows them to find their way independent of the direction they are coming from. It is light, can easily be transported and distributed while the relation between points on map and in local situation remain immutable. Latour uses the term 'immutable mobiles' for inscriptions or translations through which an entity becomes materialized into a different medium, such as a piece of paper, a digital document, a chart. These are always mobile, not only in the sense of transportable, but even more in the sense that they allow further translations and articulations in still other media while keeping some types of relations intact. Aligning such immutable mobiles, such as maps, charts, statistics etc. in a clever way allows for the production of a circulating reference, of a cascade of transformations that allows the transfer of information.[9]

Many of such immutable mobiles are present in the navigation system that Hutchins describes. These allow for 'the propagation of representations through various media' that Hutchins sees as what cognition is all about. These media are coordinated, Latour notes in his review, "by a very lightly equipped human subject working in a group, inside a culture, with many artifacts and who might have internalized some parts of the process" (Ref. 6, p. 56). Both argue for a shift in attention from mental and individual activities to trajectories of modified representations. For Latour, Hutchins' work even supplies science studies with a richer notion of mediation by artifacts, breaking away from a view on technology as "situated 'in between' mental events" and as an 'implementation' of cognition towards a recognition that "technology was the real stuff cognition was made of," that is, "[h]aving cognition is devising intellectual technologies" (Ref. 6, p. 56–57). Similarly Hutchins writes:

"these mediating technologies do not stand between the user and the task. Rather, they stand with the user as resources used in the regulation of behavior in such a way that the propagation of representational state that implements the computation can take place." (Ref. 5, p. 154)

From the previous discussion of La Pérouse's cartography one can understand that the person involved in map construction and map reading is a very lightly equipped human subject in the sense of being equipped with skills of drawing, measuring, reading maps and the like, but unrecognized in his embodied knowledge and other forms of enskilment. Hutchins found that propagation through a different medium *modifies the distribution of the required skills* and allows for simpler cognitive processes to be brought in. This last notion of a redistribution of skills allows Hutchins to extend his definition of cognition to internal phenomena, going beyond the Vygotskian notion of 'internalisation'. That is, rather than referring to something moving across a boundary, going from the outside to the inside, internalization now appears as "a gradual propagation of organized functional properties across a set of malleable media" (Ref. 6, p. 57).

However, as Latour argues for science-in-the-making, ship navigation as described by Hutchins has to be understood as a particular practice and cannot be taken as unproblematic starting point and point of comparison for understanding the rest of human activity. The situation Hutchins describes is one of the most disciplined, formalized, structured and routinized human cultures there is. Latour reproaches Hutchins insufficient attention for science studies and for the specificity of formal systems and technologies that involve immutable mobiles and transfer of information. As he points out in his review, immutable mobiles are typical of very few cognitive trajectories. The situation Hutchins describes and discusses is so specific that no general conclusions about human beings can be drawn from it. But Hutchins slips back into some old and unhelpful habits. The fact that he does not realize that his book leaves nothing standing that is essential for the existence of psychology is for Latour a sign that Hutchins is still on the fence.

"Distribution [...] does not go all the way [...] Hutchins still hesitates between a menso-centrism that would pit a mind, albeit redistributed, against a world which is simply there, and a truly symmetric anthropology, symmetry meaning not only that between Trobrianders and Americans, but that between the world and cognition."

To reconcile map and world, both map and world are constructed and "[a]s many mediations are required to transform the world into a map-like or a map-compatible shape as they are inside the ship and then inside the heads of the calculators" (Ref. 6, p. 60). The world of cartographic map users is quite different from those who go about their daily business without the use of such maps. Even more, as Ingold points out, to redistribute skill through a formal system in such a way that a break up of perception and action results is a rather rare situation that only rather recently proliferated.

2.3. *The Other Side of the Coin: Persons in a Relational World*

For Ingold Hutchins' study has one major problem. Not recognizing the specificity of the formalized system of ship navigation he is studying, Hutchins sees all of us as navigators in our daily lives, performing all the time this cognitive task as we find our way about. Though Hutchins points out the original erasure of the person as person in cognitive science he doesn't see this as part of a larger constellation that goes under the name of 'technology'. As a consequence, his generalizations on the basis of this study are one more example of disregard for the specificity of technology. For Ingold, the notion of technology is strongly related to an important historical shift in the last few centuries in which through a process of externalisation or disembedding the skilled manipulation of tools has given way to the operation of mechanically determined systems. Computation, navigation, cartography all are part of this broader phenomenon of a transfer of operations from the human body to an artificial machine or system. Latour's description of La Pérouse's mapping exercise can be considered an example in case, Hutchins' navigation system is another. And parallel to Hutchins and Latour's views, Ingold points out that such externalization does not imply a shift of content from an interior to an exterior but a rearrangement of the relations between all elements of organism and environment involved in such a way that perception and action are distanced. Whereas previously the craftsman felt the clay and saw it taking shape through the movements of his hands, the operator now sees clay going in and pottery coming out while his own actions are narrowed down to pushing a button.

But Ingold also demands attention for another aspect in Hutchins' study. The whole system of ship navigation involves quite specialized skills that can only be acquired through lengthy training and hands-on experience.

More than Latour, Ingold emphasizes that working with maps involves a skill of imagining the world as object:

> "To reconcile the chart with the territory, as Hutchins explains, one has to imagine how the world would look from a point of view — that of the 'bird's eye' — from which it is never actually seen, save from an aircraft or satellite." (Ref. 10, p. 236)

To use maps for navigation asks for certain skills and through their constant use of maps and charts the navigators become so used to thinking about the movement of the ship from this particular bird's eye angle that in their imagination the ship goes over the chart, simultaneously forgetting almost completely their on-board experience of motion. One can wonder how much this bird's eye view has become part of our daily life. This image of the globe, Ingold writes, is familiar to all of us who have gone through a western schooling and were familiarized with maps and models from quite early on. We have no difficulty imagining the world as a globe and thus as an object of contemplation, detached from the domain of lived experience (Ref. 4, p. 32). Rather, like the navigators, we might have developed a difficulty in imagining our lived-in world without reconstructing it from a bird's eye view. As Ingold notes, the separation between nature and culture, between the world-in-itself and our imaginings, "as domains respectively of matter and mind that humans in their activities must perforce seek to bridge, far from existing *ab initio*, is a *consequence* of disengagement, of the turning of attention, in thought, reflexively inwards on the self rather than outwards on the world." But that imagining of the world, Ingold points out, "is not a necessary prelude to our contact with reality, but rather an epilogue, and an optional one at that" (Ref. 11, p. 117–118). The world has become in many ways an object to handle, to confront, to walk upon from point a to b and through littering our environment with maps and globes, with signposts and name tags, street names and addresses etc. and a schooling system in which learning to read maps is a central activity, we have constructed an environment and have focused attention such that we in this modern society have indeed become at times more like navigators than humans ever have been.

We have become enskilled in taking a global perspective, at least in talk, and, as a collectivity, we have come to identify true knowledge as the knowledge gained from this perspective. That is, we have come to privilege the knowledge we get from school by looking at model globes over the knowledge we get from life by actively participating in our surroundings.

This is not surprising in a collectivity that has progressively cut out technical from social relations through a dislodging of perception and action in 'first design, then implementation'. In this process, knowledge coming from a skilled manipulation of tools has been gradually devalued, while knowledge of the explicit kind that allows one to know a situation without any personal engagement has come to be regarded as increasingly indispensable. Technology, Ingold writes, "has forced a division between knowledge and practice, elevating the former from the practical to the discursive, and reducing the latter from creative doing or making to mere execution." It brought about a profound change in the way we think about the relation between human beings and their activity. As Ingold puts it,

> "the quality of attention that [...] inheres in the skilled practitioner's conduct [is] withdrawn from the conduct itself and concentrated in the operation of a mental constructional device (an 'intelligence'), which, on the basis of a processing of sensory inputs, is supposed to generate plans and place them 'on line' for execution. Thus thought becomes active, action passive." (Ref. 10, p. 415)

This is the odd result of very specific and historically rather recently constructed human–environment relations in which the close relationship between perception and action, crucial in enskilment and skilful activity, is skillfully broken up.

Of course, the jump from La Pérouse who for the job given to him had to develop and use particular intermediaries to the building and privileging of a global perspective through schooling and building an environment full of reference points is gigantic. However, by generalizing from his particular study Hutchins shows what has made a strong proliferation of such practices possible. As many predecessors, Hutchins again develops a theory about human beings in which the privileged position of such explicit knowledge, even though distributed, was inscribed and generalized. He contributes once more to the retrojection "onto the entire field of human operations from prehistory to the present day, an understanding of the relation between practitioner, tool and material whose origin lies in the modern industrial era of machine production."[12] To undo this error, much more is necessary than broadening one's view from the individual to the sociocultural system. Hutchins' work should be the beginning of a fundamental redefinition of cognition that should be resituated in human history as part of a very recent process of externalization. Cognition, then, involving a group of people, instruments, rules, hierarchies, etc. and thus clearly socio-cultural, is also an activity in which, as Widelok notes for the Global Positioning System, "the experiental aspects of the humans involved have been systematically eliminated to leave nothing but formalized, de-personalized procedures" (Ref. 10, p. 430). To avoid a problematic generalization of this specific practice, thinking with the hands should be released from its grip and the opposition between technicity and intelligence should be undone. Dexterity and enskilment, which got lost in machine performance through the decoupling of action and perception should regain a central position in theorizing about human beings.

Ingold demands attention for the fact

> "in accord with Hutchins's ethnography but contrary to his general claim, that we are no more navigators in our everyday lives — finding our way around in a familiar environment — than we are cartographers when we retrace these movements in narrative. Navigation (or map-*using*) is [...] as strange to the ordinary practices of wayfinding as is cartography (or map*making*) to ordinary practices of mapping." (Ref. 10, p. 236)

To study these ordinary practices as a variation of cartography, navigation or computation, and thus only differing in degree from such activity is fundamentally problematic. It conceptualizes life and learning by actively participating in our surroundings as a mind versus world situation that cuts human reality at the wrong joints. Life and world are torn apart in a first instance only to be somehow artificially reconnected as mind versus world. For Ingold, signs of this original elimination of the experiential aspects of humans can be found everywhere in scientific theories about human beings. "The idea that the world exists prior to the forms of life that come to occupy it, and hence that each of these life forms is itself separately encoded in a context-free vehicle, a kind of free-floating capsule that can carry form from one site of occupation to another, is deeply entrenched in both biological and anthropological thought," (Ref. 4, p. 39) the first centralizing around the gene, the second around beliefs and representations. It should not come as a surprise then, that Ingold sees the notion of worldview involved in this mess. As soon as life and world are thorn apart into a mind versus world, it is only a small step towards filling this mind with mental representations functioning as a map to navigate the landscape and to assign meaning to a meaningless world. The fact that these latter practices are based on an elimination of human/personal input in order to gain (a constructed) universal validity, is forgotten by taking the results, the maps

152

and charts, disconnected from their construction processes, as a starting point and standard in the study of human beings. Not surprisingly, these appear as beings carrying concepts with them while moving around in the world, rather like carrying a map in navigating the landscape. "In different times and places I experience different sensations, but like a map, the system of concepts which organizes these sensations into meaningful patterns remains the same regardless of where I stand" (Ref. 10, p. 408). Blindness for the specificity of the activity of erasing the human in computation and navigation results not only in a separation of mind and world, but also in the construction of a non-human human, of a detached from the world homunculus, capable, through context-independent structures and properties, of thinking and learning about the world prior to bodily engagement with it. A relational context of being-in-the-world is turned 'outside in' to become a cognitive attribute of the mind.

3. Worldview as a Non-Relational Notion

From the previous discussion it becomes clear that many concepts need to be reconsidered. Culture is one of these, worldview is another. For Ingold, the notion is highly problematic. The lifeworlds of people of different cultures are often treated "as alternative constructions, cosmologies or 'worldviews', superimposed upon the 'real' reality of nature" (Ref. 10, p. 15). A worldview is then seen as a particular cultural construction of an external reality. It stands as culture opposite to nature, and both concepts relate to a world existing "for a being that does not belong there, and that can look upon it, in the manner of the detached scientist, from such a safe distance that it is easy to connive in the illusion that it is unaffected by his presence" (Ref. 10, p. 20). Bridging the distance by transferring some of this culture or worldview into material artifacts and objects figuring as vehicles or carriers of meanings, distributing it across heads and inscribed objects, does not really solve the original error of a priori detaching the mind from the world. Such a mind has to formulate a world prior to any attempt at engagement. The assumption is that through a process of mental representation a view is constructed of the world. For Ingold, however, "apprehending the world is not a matter of construction but of engagement, not of making a view of the world but of taking up a view *in* it" (Ref. 10, p. 42). In *Key Debates in Anthropology* Ingold argues against the view that human worlds are culturally constructed. In fact, what he rejects is "the very conceptual paradigm that forces us [...] to choose between animality and humanity, between one world and many worlds, between nature and culture, and between the given and the constructed" for the reason that "these four dichotomies are linked by a common, cognitivist orientation that contrives to disembed individual human beings from the relational matrix of their existence in the world, only to re-embed this world inside their individual heads" (Ref. 11, p. 117, 112). Considering a worldview as a necessary attribute that allows human beings to navigate in their world is for Ingold making the mistake of attributing "to native people of a sense of what it means to know one's whereabouts that effectively treats them as strangers in their own country" (Ref. 10, p. 219). La Pérouse's relation to the local situation which for Latour contained the explanation for the difference between cartographic knowledge and the locals' embodied knowledge of their surroundings is projected upon the latter. Worldview as a non-relational notion thus contains and generalizes the separation of perception and action: a 'worldview' is then "a set of rules and representations for the organization of sensory experience that individuals carry in their heads and that are available for transmission independently of their bodily activity in the world" (Ref. 10, p. 225).

The problematic nature of worldview as a non-relational notion becomes obvious in Sire's discussion of the concept. Having a worldview is simply taken as a starting point. Summarizing Naugle's history of the concept, he notes the close connection with a human mind or human consciousness versus the external world perspective. It is this assumption that makes the rejection of the notion self-refuting, Sire notes, since everyone, relativists included, make statements about reality (Ref. 1, p. 29). One can wonder of course if reality is understood here in a globe perspective or a dwelt-in world perspective. It seems obvious though that once a mind versus world perspective is taken as starting point that 'a worldview' easily receives the characteristics of a global map that stands in a referential, cartographic relation towards 'reality-as-it-is-in-itself'. Furthermore, as a post-Kantian notion, worldview further involves the assumption that "what a person perceives is primarily dependent on the mind of the perceiver [...] we understand that reality through the structures inherent in our own mind" (Ref. 1, p. 26). For Sire it is obvious that "a worldview is the shaping structure of our own autonomous selves. We see what *we* see." He refers to Wilhelm Dilthey for whom the basic role of a worldview is "to present the relationship of the human mind to the riddle of the world and life" (Ref. 1, p. 25). It takes shape through a complex interrelation between human consciousness and the external world that results in a more sophisticated and detailed sense of who we are and what is around us. For Dilthey, then, there is a common human nature and a common reality, but one's worldview is one's own, developed under specific conditions of climate, race, nationality, determined by history, through political organization etc. A worldview is seen as the bridge between a hopelessly lost and confused human mind and the world, giving this mind a stability, "a solution to the riddles of life that provides a way of successfully thinking and acting in the world" (Ref. 1, p. 26). According to Sire everyone needs answers to questions such as "what is prime reality?", "what is the nature of external reality?", "what is a human being", "What happens to a person at death", "Why is it possible to know anything at all?", "How do we know what is right and wrong?" and "What is the meaning of human history?". Answers to these questions are needed since, as he sees it, these attempt to get at the foundations of all our human thoughts and actions (Ref. 1, p. 94). However, it is obvious that this series of questions are put in a universal phrasing and expect universal answers, answers that can give anyone anywhere an idea about the world, about him/herself and about what to do regardless of where they stand. The cartographic map seems like a good metaphor to clarify his views. What is obvious is that the human being asking such questions and needing those answers stands in a particular relation to 'reality-as-it-is', rather than finding him or herself within an already dwelt-in world. It seems then that the exercise of worldview construction, as Sire envisages it, carries in itself a relation of disengagement and thus results not surprisingly time and again into a universalization of having a worldview to reconnect what has been torn apart. As a consequence, the exercise involves the projection of the basic scheme that is built into the questions onto the subject matter 'human beings'. This shows in Sire's view that everyone has a worldview, whether they are conscious of this fact or not. Talk about the pretheoretical nature of worldviews and the innateness of basic concepts are easy means to make the claim stick, but new problems arise.

"A worldview is simply inescapable" (Ref. 1, p. 112–113, 121). Inescapable as it is, Sire nevertheless wants to acknowledge the commitment that some worldview owners consciously make. At this level, he still is confronted with the difficulty that apparently some of these worldview owners make a commitment to a worldview positing a God or gods or other things which lack public proof. The case he makes is simple. Our worldview is presuppositional, Sire explains, since in the end one cannot prove a worldview beyond the shadow of a doubt and consequently, our commitment to it is at least in part a matter of faith. Sire wants to distinguish presuppositional from pretheoretical. The latter refers to that without which we cannot think at all, what is so intuited that we can't think any other than the way we do. It consists of the notions and recognitions of their inter-relations that precede any thought at all. "It is what we think with, not what we think about" (Ref. 1, p. 79). In this sense, the concept of good is for Sire a universal pretheoretical given, part of everyone's innate, initial constitution as a human being. The presuppositional character of a world-view on the other hand lies in the fact that though reasons can be given, proof is absent. Nevertheless a commitment is made and one lives one's life in accord with it since it allows us to give enough meaning to our life to keep going. An answer to what happens after death for instance is a belief held on faith. Sire finally explains the difference between pretheoretical and presuppositional by considering the first as the bones, the latter as the flesh of worldviews (Ref. 1, p. 86).

The problem that arises from distinguishing the pretheoretical and pre-suppositional character of a worldview is how these are related. Sire's metaphor of flesh and bones hardly does the trick. But the whole pic-ture is further complicated through the introduction of a new dichotomy of the private and public dimensions of worldview. Worldviews, Sire notes, "are both specific commitments held by individuals and sets of assump-tions that characterize a specific community, historical era or entire cul-ture" (Ref. 1, p. 107). It would seem then that the individual addresses the seven fundamental questions but does so on the basis of a set of as-sumptions shared with others. However, when Sire addresses the private side, things seem to get mixed up. "Whether we know it or not, we all operate from a set of assumptions about the world that remain to a large measure hidden in the unconscious recesses of our mind. That worldview is private." Such worldview seems largely implicit as Sire writes that "in some way all of the basic worldview questions are being answered by the way I am acting and behaving." At such a level it seems that answers to worldview questions do not precede acting in the world, rather the acting itself provides the answers. Worldview in its private dimension is then an integral part of who one is, "[i]t will be so much a part of what is uniquely me that there will be no other worldview in the universe that is identical to my own" (Ref. 1, p. 108). Not surprisingly, Sire ends up with a new problem. The question now is "if worldviews are so private and unique, how can they characterize a community or a culture?" Some unity of ideas, a shared map, seems necessary. Predictably, to answer the previous ques-tion, 'human nature' is seen as a good candidate. We are individuals but also members of the human family with whom we share some basic categories. In a second step, things like cultures, religions, social class etc. come in to point to groups of people with whom we share some presuppositions. Those groups become smaller and smaller until finally the individual level is reached again. However, such a story connecting private and public aspects tell the story simply in terms of shared presuppositions. Worldviews appear as containers of ideas and propositions that are transmitted either by genes, through cultural transmission or some combination of these. They appear as theories about world and life that are somehow present without any theorizing. However, Sire notes, such a cold notion of worldview, quite compatible with the cartographic map metaphor, seems to carry in itself a certain distance between people and map. Attempting to explain the strong commitment that believers can have even towards beliefs for which no public proof is present, this is an aspect Sire isn't happy with.

In his attempt to close the distance, Sire comes close to breaking up the Cartesian framework, but freezes at the last moment. As he points out, worldviews have been understood to be inextricably tied to lived experience and behavior. They are incarnated, fleshed out in actual ways of life. Worldviews are perceptual frameworks, that is they are 'ways of seeing' which as Walsh and Middleton argue can be assessed by observing how we or others act. As an example they describe how four families in four different cultures from the same social classes care for their babies. Their notion of worldview comes very close to Geertz' notion of culture: Each set of actions in this example, Sire points out, "illustrates a different 'life form' or worldview. A world view is never merely a vision *of* life. It is always a vision *for* life" (Ref. 1, p. 98). Every operative worldview directs action. However, how this goes about is never really theorized and one can ask whether life form, way of life and worldview can be so easily equated as Sire makes it appear. Sire limits himself to a discussion of worldview as master story. He points out that folklore, myths and literature from around the world present stories about human reality in the larger context of universal cosmic and human meaning and sees these acting as orienting patterns. This leads him to question the description of worldview first in intellectual terms as 'system of beliefs', 'set of presuppositions' or 'conceptual scheme'. Such a description, he claims, misses an important element in how people actually think and act and he proposes to see a worldview as "the *story* we live by." Religious practice is full of telling stories, he points out, and worldview to be understood as meaning is transmitted by these narratives. Such stories are not just accepted in the form of intellectual assent, for Sire, "[w]e do not just believe it *at a distance. We are to indwell it as if it were our own story, because it actually is*" (Ref. 1, p. 15, italics added). Is Sire here trying to overcome the disengagement and estrangement that such talk about reality-in-itself and worldview as (cartographic) map entails? He quotes Newbigin to underline his point:

> "The Christian community is invited to *indwell* the story, *tacitly* aware of it as shaping the way we understand, but *focally* attending to the world we live in so that we are able confidently, though not infallibly, to increase our understanding of it and our ability to cope with it." (Ref. 1, p. 105)

This quotation makes it clear that these Christian thinkers on worldview swing from a strongly Cartesian view to something that comes close to resembling the alternative view that places processes of enskilment at the centre. The 'master story' becomes what educates the attention. "To indwell a story is to live so much within its framework that we are not so conscious of the story as of what the story allows us to see." The story becomes a means that allows people to see, like a telescope allows people to see without them really seeing the telescope. "Rather we *indwell* the instrument in order to do what we could not otherwise do, to see things we otherwise could not see" (Ref. 1, p. 105). One consequence of seeing worldview as a means is that it becomes obvious that thinking about one's worldview is a rather rare and specific activity. Normally we do not think about a worldview, "rather we *think* with our worldview and *because* of our worldview, not *about* our worldview" (Ref. 1, p. 124). Only at specific moments, people engage in the latter activity. However, where Sire sees the story as putting us in contact with the way God, the world and we really are and emphasizes over and over again the primacy of ontology over epistemology, a symmetrical anthropology allows a view that breaks away from viewing worldview as something between us and the world — Newbigin for instance sees worldview as "our 'reading glasses', 'our telescope', our 'place to stand' to view reality, the hub of our world, the heart of our selves," all, except maybe for the last two, metaphors in which vision or the observer stance is still preferred and easily universalized. Sire's characterization of worldview as master story is interesting, however, Ingold shows that paying attention to the activity of storytelling brings a whole different perspective on things.

4. Worldview as Relational Notion?

In a chapter "To journey along a way of life", Ingold wonders about the difference between the map-using stranger who finds his way in unfamiliar country through identifying his or her position with a certain spatial or geographic location, defined by the intersection of particular coordinates on a map, and the person who has grown up in that country going about without an artefactual map. Rather than simply projecting the map into the local person's brain or mind, Ingold notices that ordinary wayfinding and mapping more closely resemble storytelling than map-using (Ref. 10, p. 220). The reason is that one knows *as* one goes and through storytelling and narrative gesture the experience of moving from place to place in a region is re-enacted. Sometimes this leaves traces in sand, sometimes traces of a more lasting nature. As Latour would confirm, this primordiality of movement even goes for scientists and explorers such as La Pérouse who necessarily had to sail to the Caribbean in order to map the area. However, Ingold points out, it is one of the most striking characteristics of the modern map that it erases the practices and itineraries that contributed to its production (Ref. 10, p. 230). To use such a map or such a truncated view on maps as the central metaphor in theorizing human beings is to forget the primacy of such practices and itineraries. People know as they go, not before they go, in some 'pretheoretical' moment.

> "Knowledge of the world is gained by moving about in it, exploring it, attending to it, ever alert to the signs by which it is revealed. Learning to see, then, is a matter not of acquiring schemata for mentally constructing the environment but of acquiring the skills for direct perceptual engagement with its constituents, human and non-human, animate and inanimate." (Ref. 10, p. 55)

Such a view also implies that life is not simply the time spend by an organism *on* this world as good or as bad as it goes with the tools it has at hand. Rather, itineraries and practices, or life activities, are generative of personhood. Life in this perspective is a process in which every being arises, is caught up while carrying it forward. Ways or forms of life are then not scripts transmitted from mind to mind to put in practice, but are histories, processes "wherein the people of each generation furnish through their life-activities the contexts within which their successors grow to maturity" (Ref. 10, p. 384). One comes into existence and undergoes continual formation throughout one's life within a wider field of social and environmental relationships while simultaneously contributing through one's activity to the perpetuation and transformation of that field. Becoming a person, then, is laying down a trail, a life-line, a trajectory of change and development. Growing is "moving along *a way of life*, conceived, not as the enactment of role-specifications received from predecessors, but literally as the negotiation of a path through the world."[13] The movement within this field of continuously formed and forming relationships, then, is not only the source of people's perceptual orientations, but also of their very substance as being. Furthermore, this movement involves a reciprocal contribution to the substantive make-up of others as well. In these life processes cultural forms appear, such as basket weaving or brick laying or meditation, which are an aspect of the growth and development of persons, involving repeated tasks and exercises through which novices become experts. Such learning processes do not involve a transmission of cultural information but involves the alignment of one's movements through observation and a focusing of attention with that of the expert. Stories and artifacts, as re-enactments, condensations or sedimentations of movements, thus not necessarily simply concern one's travels in the world but also the movement of becoming a person. And just as some lines on a paper are but the steppingstones along the way punctuating the narrated process of someone's travels, so also instructions are rather steppingstones or signposts which get meanings from their positioning within a field of practice.

From such a perspective on storytelling, movement and life the following questions can be posed. If traditions are to be seen as constantly changing bodies of temporary and less temporary intermediaries, of signposts along a path of life, might not then what Sire describes as the presuppositional nature of worldviews in many cases possibly be better understood as a sustained commitment to walk a certain life path, walked by others and with some of these others, actively participating in a tradition that is thus carried forward? Furthermore, one can wonder whether differences here, between traditions, more particularly between so called small traditions and world religions, might not also be a consequence of movements and goals. Are world religions or more systematized traditions with scriptures and doctrines examples of more elaborate and stabilized intermediaries that allow their transportation over long distances? Talking from the Christian tradition, Latour describes religion as, like science, a form of displacement but then notes an important difference. Whereas the sciences are to be seen as a form of displacement through transformation and information, religion is a form of displacement through transformation and translation. Both involve a cascade of transformations, but the goal is different. In the case of science, circulating references are constructed to transport information, in the case of religion, Latour writes, "containers that make persons emerge" are transported.[14] Religious talk or religious images should not be decrypted as if it transported a message but as if it transformed the messengers themselves. Illuminating the Christian regime of invisibility through a discussion of the path traced by religious images, he points out that the purpose was not to turn the spectator's eyes to something in the past as if the image was transferring information about someone living there and then, but that "on the contrary, incredible pain has been taken to *break* the habitual gaze of the viewer so as to attract his or her attention to the *present* state [...] the idea is to avoid as much as possible the normal transport of messages, [...] [t]he aim is not to add an invisible world to the visible one, but to distort, to opacify the visible world enough, so that one is not led to misunderstand the Scriptures but to re-enact them truthfully."[14] Might the story of the camel and elephant sustaining the world not be a similar case?

However, whereas traditions might be seen as a continuously changing constellation of intermediaries, as means that focus attention and as surroundings grown over time in a way of life that is continuously advanced through the activity of its participants, the intermediaries one would construct when one, as La Pérouse, has no intention to participate, to stay and to become enskilled in that tradition will likely take a different form. Until now, one can say, this has mainly taken the form of mapping out 'the culture' or 'the worldview' of the other. Anthropologists have been very productive here. But such contacts, such crossing of other people's paths, has increased exponentially in our globalized world and it would seem to me that a study of such worldview construction, the whens, the wheres, the whys and hows is highly necessary. One discipline that comes to mind for such a study is religious studies. So far, this discipline has been mostly involved in the study and even the construction of religious worldviews in order to compare them on the basis of some basic themes. Interestingly, since the seventies and eighties, the globalisation theme has become quite central. Views of a 'global village' or 'global city', of a strongly interconnected world with increased local religious diversity and increased inter-religious and inter-cultural contact form the background for new methodological and theoretical proposals. As Ninian Smart noted (Ref. 15, p. 418), cultural interaction is inescapable now, as is the need to adjust to the new secular forces both inside and outside the West of modern science, industry, capitalism, individualism and so on. He sees in this Global Period a trend in which religious traditions see themselves more and more as world religions, resulting in a new feeling of identity. In the last 200 years, Smart says, a new religious mode came into existence: religion aware of itself. That is, contact with major world cultural forces and, especially, contact with secular values and with universal proselytizing faiths like Christianity compels traditions to place themselves in the same league of universal talk, shared with non-religious worldviews. So far this phenomenon has been studied mainly as a clash of worldviews, with worldview taken as a relation-blind notion. Naugle and Sire's work are examples of such work. But, since they obviously write from an engagement towards the Christian tradition, one can also wonder, how much the participants in these trends, working to consolidate their traditions in the context of religious pluralism are caught into that same non-relational way of thinking. Asking and answering universal questions about life, about the world, about oneself drawing out a map that is considered useful regardless of one's personal situation and relational environment is problematic especially if as part of the universal message its universality is legitimized through the proposition that life generally and necessarily takes the form of asking these questions and longing for these answers.

What happens if we consider worldview as a relational term, as an intermediary rather than a thing in one's mind or society? If we avoid the mistake of ignoring movements, itineraries and practices by inscribing in theory as well as in worldviews their own universality and primacy, what might worldview construction involve? When would it be useful? What if we start to understand worldview inside out instead of outside in, as a translation not of an in-the-mind thing but as traces or externalizations of organism–environment relations characterized by direct perception and enskilment processes? Might it be useful to exchange the term worldview for world perspective, if the latter notion not only stands for sight but for feel, for smell, and even more for enskilled perceptual engagement with one's surroundings? A symmetrical anthropology of the modern should be developed in which worldview construction as a reality now is mapped out. Is worldview construction and dialogue together with world religion talk and interreligious dialogue just a big industry, as some have written, creating a need in the costumer for which they subsequently provide solutions? Such a debunking view is in my estimation a too easy accusation without a proper trial. Possibly, worldview construction and worldview use appear as in a sense comparable to ship navigation, as a form of distributed cognition that extensively builds upon charts and maps, books and documents

162

that, similar to the charts and other media aboard the Navy ship, modify the distribution of involved skills. As Hutchins and Latour argued for the study of cognition, likewise a shift from mental and individual activities to trajectories of intermediaries and translations might throw a quite different light on worldview construction and reflection. It seems to me that we have here an important subject matter for study, the more so since worldview construction and reflection are often mentioned when important problems in our globalized world are dealt with. Quite often contents of worldviews are taken to be the necessary focus looking for possible compatibilities, complementarities or conflicting truth- and practice-claims. But, if worldview construction, reflection and use is rather a quite specific practice involving intermediaries that involved a disengagement between person and world, we should also consider the possibility that conflicts might be a consequence of a mistaken generalization of this situation through simply locating worldviews in people's minds.

Acknowledgments

The author is Postdoctoral Fellow of the Research Foundation–Flanders (FWO).

References

1. J. Sire, *Naming the Elephant. Worldview as a Concept.* Downers Grove, Il.: Intervarsity Press (2004).
2. D. Aerts, B. D'Hooghe and N. Note, *Worldviews, Science and Us. Redemarcating Knowledge and Its Social and Ethical Implications.* New Jersey: World Scientific Publishing, 1 (2005).
3. G. M. Lewis (Ed.), *Cartographic Encounters: Perspectives on Native American Mapmaking and Map Use.* Chicago: University of Chicago Press (1998).
4. T. Ingold, Globes and Spheres. The Topology of Environmentalism. In: K. Milton (Ed.), *Environmentalism. The view from Anthropology.* London: Routledge (1993).
5. E. Hutchins, *Cognition in the Wild.* Cambridge, MA: MIT Press (1995).
6. B. Latour, Cogito Ergo Sumus! Or Psychology swept inside out by the Fresh Air of the Upper Deck... *Mind, Culture, and Activity: An International Journal* **3** (1) (1996).
7. B. Latour, *Science in Action. How to Follow Scientists and Engineers through Society.* Cambridge, MA: Harvard University Press (1987).
8. B. Latour, Comment redistribuer le Grand Partage? *Revue de Synthèse* **110**, 224 (1983).
9. B. Latour, *Pandora's Hope. Essays on the Reality of Science Studies.* Cambridge, MA: Harvard University Press (1999).

10. T. Ingold, *The Perception of the Environment. Essays in Livelihood, Dwelling and Skill*. London: Routledge (2000).

11. T. Ingold, *Key Debates in Anthropology*. London: Routledge (1996).

12. T. Ingold, 'Tools for the Hand, Language for the Face': An Appreciation of Leroi-Gourhan's Gesture and Speech. *Studies in the History and Philosophy of Biology and the Biomedical Sciences* **30** (4), 432 (1999).

13. T. Ingold, Between Evolution and History: Biology, Culture, and the Myth of Human Origins. *Proceedings of the British Academy* **112**, 51 (2002).

14. B. Latour, 'Thou shall not Freeze-Frame,' or, How not to Misunderstand the Science and Religion Debate. In: J. D. Proctor (Ed.), *Science, religion, and the human experience*. Oxford: Oxford University Press (2005).

15. N. Smart, A Religious Worldview for the 21th Century. In: *Cosmos, Life, Religion. Beyond Humanism. Tenri International Symposium 86*. Tenri, Japan: Tenri University Press, 418 (1988).

THE STRUCTURES OF KNOWLEDGE
IN A WORLD IN TRANSITION

RICHARD E. LEE

Director, Fernand Braudel Center
Binghamton University
Binghamton, NY 13902-6000
E-mail: rlee@Binghamton.edu

The structures of knowledge, the separation of facts from values into the two cultures of what eventually would come to be called the sciences and the humanities, emerged as fundamental components of the modern world-system along with the axial division of labor and the interstate system in the long sixteenth century in Europe. The restructuring of the late nineteenth century resulted in the creation of the social sciences between the poles of the sciences and the humanities. This structure is now in crisis. Developments in complexity studies and cultural studies put into question not only the utility of the presently accepted disciplinary boundaries, but as well the attendant received epistemological, methodological, and theoretical approaches.

1. Introduction

Half a century ago, during the post-1945 period through the mid-1960's say, it was quite clear to academics and laymen alike what the organizing arrangements of the disciplines of knowledge, and the layout of the departments that expressed them institutionally, were. At the upper end of what was indeed a hierarchy of disciplines were the natural sciences; at the bottom of the hierarchy were to be found the humanities; and somewhere in a contested, amorphous, middle space were the social sciences. The departments of the superdisciplines were generally housed in different buildings on university campuses; libraries of the sciences were often separated from those serving the humanities and the social sciences; and the departments themselves published in different sets of proprietary journals.

The unquestioned prestige of the sciences was premised on what were genuine accomplishments in explaining real world phenomena and thus at least to a certain extent harnessing those phenomena, as for instance witnessed in the war effort (radar, atom bomb, etc.). There was also, however, an epistemological dimension to this prestige. The sciences purportedly dealt with the natural world in terms of universal truths, in terms of 'facts', which could either be observational (specifics) or theoretical (natural laws). The key term was universal in the sense of always, everywhere and untainted by human bias. Furthermore, scientific laws were constructed in such a way as to admit prediction. The humanities, on the other hand, in considering the world of human relations proposed anything but universal knowledge; they were relativistic and explicitly concerned with the unpredictable arena of human values. Finally the social sciences, which were also concerned with the world of human relations, sought to develop a form of knowledge about the social world that was 'scientific' in the sense of being unbiased (the 'view from nowhere') and admitting at least a soft form of predictability. The story of how this structure, the structures of knowledge, developed, and how it had been and continues to be articulated with the material world of economic and political processes is also the story of the modern world-system.

In the simplest terms, by the structures of knowledge I mean those patterns of what can and cannot be thought that determine what actions can and cannot be deemed feasible in the material world. These modes of cognition and intentionality are co-constitutive of the modern world-system with the (economic) structures of production and distribution in the form of the axial division of labor and the (political) structures of coercion and decision-making manifested in the interstate system. They are unique in human history in that they admitted and admit two ways of knowing, rather than just one. Those two ways of knowing were premised on a differentiation and separation between 'truth' and 'values' and were eventually institutionalized as a hierarchical separation between the 'sciences' and the 'humanities', that is, the 'two cultures'.[1]

The medieval structures of knowledge recognized diverse fields or subject-matters; rhetoric was certainly not astronomy. What was not recognized was differing bodies of knowledge that were based on contradictory visions of the way the world worked. It was this new epistemological divide, and the hierarchy that privileged as legitimate and authoritative 'factual knowledge', that would become the norm. The mechanisms of this transformation, such as the establishment of double entry bookkeeping (see Ref. 2) and the renewal of the authority of the principle of the excluded middle, were clearly articulated with the political and economic developments we generally consider the markers of what is termed the transition from feudalism to capitalism. These markers included the rise in the status of merchants and the legitimation of profit making (the *virtù* of the balance in the 'bottom line'), the development and integration of the world market, and the transformation of the basic political entities of 'realms' based on 'parcellized sovereignty' into 'states' with borders.

The long-term trend deepening this structure underwent two great conjunctural adjustments or 'logistics' analogous to the waves of expansion and contraction in the economic arena and the cycles of relative concentration of power in the geopolitical realm.[3] The first consisted of the seventeenth-century Newtonian synthesis between Baconian induction and empiricism and Cartesian deduction and rationalism, which created the foundation for the dominant theoretical approaches and methodological practices in the sciences and led to the solidification of the separation of the sciences from the humanities.

The second was the late nineteenth-century creation of the social sciences, situated between the sciences and the humanities. The putatively value-neutral social sciences, which seemed to offer the possibility of a 'scientific' or non-value-oriented policy-making process in the service of 'progress', came to occupy a tension-charged space in the wake of the irresolvable contest between the equally value-laden but mutually exclusive positions taken by conservatives (tradition, order) and radicals (democracy, anarchy) in the humanities on the political future of the world following the French Revolution. The key controversies in this process were the late-nineteenth-century *Methodenstreit* in the German-speaking world (in both philosophy and economic history/economics: Dilthey and the Baden neo-Kantians, Schmoller, Menger) and the English order and anarchy debates (Carlyle, Arnold, Mill) — the one taking place in the context of state-formation and development in the Germanies, the other in the context of British political unrest at home and uprisings in the colonies.

Instead of value sets, on which the politics of both the right and the left had been founded, the social sciences increasingly ordered collective decision-making along the lines of Mill's suggestion that from the 'science of society' come 'guidance' (Ref. 4, p. 64). T. H. Huxley called on the objective, value-neutral, problem-solving spirit of science to realize progress without moralism. On the occasion of the opening of Sir Josiah Mason's Science College, Birmingham, in 1880, Huxley delivered a lecture in which he asserted that

> "if the evils which are inseparable from the good of political liberty
> are to be checked, if the perpetual oscillation of nations between

anarchy and despotism is to be replaced by the steady march of self-restraining freedom; it will be because men will gradually bring themselves to deal with political, as they now deal with scientific questions." (Ref. 5, p. 158–159)

The political consequences of this medium-term solution were the 'scientific' legitimation of the association of particular 'nations' or 'peoples' with individual states, the hierarchical placement of groups on a racialized and gendered world division of labor, and the rise of the 'new liberalism' effectively eliminating clear alternative political agendas on the right and on the left.

Thus, this division of labor found on one side the factual, universal, positivistic, empirical, objective, fact-producing, and quantitative disciplines of the sciences engaged in explaining order in a world where past determined a predictable future via universal laws. On the other side was to be found the particularistic (for instance, with regard to social contexts, locales, or time frames), chaotic, value-oriented, and qualitative disciplines of the humanities where scholars dealt with an unpredictable and relativistic world of free human agency. In their quest for legitimacy after 1945, the social sciences deepened their efforts to emulate the putative universalism of the natural sciences. They were, nonetheless, divided on questions of both theory and method. On the one hand, universalism was expressed in quantification and the comparative method in economics (econometrics), sociology (structural-functionalism), and political science (behaviorism), while, on the other hand, universalism was expressed additively in the more narrative bent of history and anthropology. Although all the disciplines exhibited to some extent both tendencies, scientism seemed to be gaining throughout. Similarly, even the humanities, in their effort to retain a credible voice, sought to echo the decontextualization, atemporality, and presumptive objectivity of the sciences with approaches such as 'new criticism' and 'close reading'.

The acceptance of this structure, and its hierarchy, as natural and beyond question reached its peak in the immediate post-1945 period, and, like the axial division of labor and the interstate system, it became global in extent. Nonetheless, tensions have remained apparent over the past half century in the resurgences of old epistemological debates such as reductionism versus holism, structure or determinism versus agency or freedom, and order versus disorder — each antinomy manifesting an aspect of the division between facts and values, the sciences and the humanities.

Today, there is certainly a heightened sense that we live in a 'knowledge-based society'. There is, indeed, much discussion of this in the standard literature. Although it is quite true that analysts and entrepreneurs alike have always considered knowledge to be a central aspect of the processes through which capital is accumulated, it is only since the 1960's that the realm of knowledge has been in any way problematized, even in the short run. Fritz Machlup initiated research in this vein by examining 'knowledge industries',[6] apart from the rest of the service sector of the U.S. economy, and measuring their contribution to the gross national product. He concluded that they would soon overtake the industrial sector. Since then, both the scholarly and popular literature on the 'information society' and the 'knowledge society' has exploded.

Today, there are those who suggest that we are living in an age of a 'knowledge economy' (e.g., Refs. 7–9). Others suggest that we are living in an age of 'knowledge capitalism'. For his part, Lester Thurow considers the twentieth century as a period of transformation from a natural resource economy to an economy in which the major players are brainpower industries and in which the major sources of comparative advantage are knowledge and skill alone.[10] To cite another particularly well-known observer and analyst, Peter Drucker contends that today we are living through a transformation to a post-capitalist society. He notes that "knowledge is becoming the sole factor of production, sidelining both capital and labor" (Ref. 11, p. 20).

I would argue that the transformation is much more fundamental and profound than either of these or many other mainstream analysts suggest; it is the working out of the secular crisis of the processes reproducing the structures of social life of the modern world that emerged from the transition from feudalism to capitalism in Europe. However, it is only visible as a fundamental crisis from the standpoint of the *longue durée*.[12] Nonetheless, the recognition of this crisis is, indeed, the essential starting-point for relevant social analysis today.

The crisis was implicit within the sciences over the two decades after 1945 when voices that diverged from the standard positivistic, reductionist and analytic model in the direction of an organic, relational model began to be heard. In one way or another, these new approaches were conceived as appropriate to the study of "systems that are intrinsically complex", as Ross Ashby noted (Ref. 13, p. 249). *Organized complexity*, the middle ground Warren Weaver posited between simplicity (the problems of classical physics with few variables, the realm of necessity) and disorganized complexity (problems with many variables amenable to statistical description, the realm of chance),[14] had proven particularly resistant to the analytic method, that is, to the development of mathematical models or equations expressing general laws in which all contributing causal factors appeared as variables. The issues involved were addressed by, among others, the elaboration of General System Theory (GST) (e.g., Ref. 15) and Cybernetics (e.g., Ref. 16).

But these have not been easy roads. In matters of application, Anatol Rapoport, an original protagonist of GST, expressed concern about the exploitation of systems approaches in the management of military organizations and in consulting work where the goal was corporate profit-making (Ref. 17, p. 16). Felix Geyer has articulated the larger question:

"the more realistic — and therefore less parsimonious — a theory, the more complex it becomes, and the more difficult to test the hypotheses and subhypotheses derived from it which are used in collecting and interpreting the data [...] For the time being, sociology should perhaps [...] force itself to give up the ambition to make accurate medium- and long-term predictions, except in delimited areas of research where complexity is still manageable or can be more or less contained." (Ref. 18, p. 28)

This statement alludes to both a boundary problem and to the ultimate ambition of the social sciences: prediction. It has become clear that defining a set of interactions for which the external context can be ignored (in the sense of having negligible impact) is extremely problematic. Furthermore, work in contemporary complexity studies shows that prediction itself is possible for only a subset of systems studied even in the natural sciences, and then only under certain, limited circumstances — much less for the world of human interactions.

Complexity studies emerged directly from research in the sciences and mathematics (see, for instance, Refs. 19–24). At the very moment of the worldwide triumph of the Newtonian worldview, that is, of a deterministic world of natural laws based on time-reversible dynamics, this new knowledge movement challenging the premises of that worldview began to take root. The rethinking that we are witnessing today represents a synthetic approach as opposed to a reductionist one; indeed, a science of complexity holds out the possibility of representing change — that is, "describing our collective reality as a process" — without reverting to reductionism (Ref. 25, p. 273). It marks a shift away from the Newtonian worldview emphasizing equilibrium and certainty and defining causality as the consistent association of antecedent conditions and subsequent events amenable to experimental replication and hypothesis testing. In 1986, Sir James Lighthill, President of the International Union of Theoretical and Applied Mechanics, went so far as to apologize on behalf of "practitioners of mechanics [...] for having misled the general educated public by spreading ideas about the determinism of systems satisfying Newton's laws of motion [implying complete predictability] that, after 1960, were to be proved incorrect" (Ref. 26, p. 38).

Although all developments were not equally successful, what did turn out to have enormous resonance, in terms of extensive theoretical pertinence and broad areas of application, was the study of chaos. The recognition of the existence of chaotic behavior exhibited by nonlinear systems flew in the face of Laplacian predictability. As these systems evolve over time, they rapidly magnify small perturbations and are thus highly sensitive to small changes in initial conditions. Despite this, there remains evidence of an embedded order underlying the seemingly random evolution of certain dynamical systems. The development of what came to be loosely known as chaos theory on so many fronts opened up the possibility of applying deterministic models, formerly restricted to the 'closed universe' of 'completely predictable systems', rather than stochastic models, to certain systems that behave randomly. Such randomness, of course, leaves open a place for chance and therefore creativity and change. But in natural systems, not all theoretically possible states turned out to be realizable. Only some, those that lie on the strange attractor of such systems, will actually appear in nature. Ivar Ekeland has called this an "admirable and subtle mix of chance and necessity" (Ref. 27, p. 13, 12, 15).

Drawing on an immense body of work by Ilya Prigogine (which won him the Nobel Prize) and carried on with his colleagues in Austin and Brussels where complexity has been understood in terms of system 'behavior' rather than of system interactions,[28] Prigogine and Isabelle Stengers presented chaos not as the opposite but as the source and confederate of order.[29] They considered that a conceptual transformation of science was taking place. This transformation was growing out of the challenge to Newtonian mechanics associated with contemporary research in thermodynamics focusing on nonlinearity (instability, fluctuations, order-out-of-chaos) and the irreversibility of the evolution of far-from-equilibrium, open systems, characterized by self-organizing processes and dissipative structures. In the light of instability and chaos, and the association of the arrow of time with order as well as disorder, Prigogine and Dean Driebe have maintained that the laws of nature now express possibilities instead of certainties. There is no longer any contradiction between dynamical and thermodynamical descriptions of nature. Far from being a measure of our ignorance, entropy expresses a fundamental property of the physical world, the existence of a broken time symmetry leading to a distinction between past and future that is both a universal property of the nature we observe and a prerequisite for the existence of life and consciousness (Ref. 30, p. 222).

Prigogine, looking for the roots of time, "became convinced that macroscopic irreversibility was the manifestation of the randomness of probabilistic processes on a microscopic scale" (Ref. 31, p. 60). The recognition that probability is more fundamental than trajectories implies what Prigogine calls 'the end of certainty',[31] and the end of certainty in scientific prediction connotes (again) an open future of creativity and choice in natural systems and, as a consequence, a vindication of freedom, agency, and creativity in our understanding of social systems, but with particular significance in times of crisis or transition.

In the humanities, it was the turn to culture, or rather the return to culture — and the emergence of cultural studies as a consolidated knowledge movement (see Ref. 32) — by the independent left in Britain during the 1950's that signaled the crisis of the structures of knowledge. And indeed, this was part of a specific historical conjuncture. The context was formed in the short term by the geopolitical events of 1956 (Hungary, Khrushchev's 'secret' speech, Suez) that figured in the East–West struggle and in the medium term by the multi faceted dominance (hegemony) of the United States. On the one hand, the category of culture had enjoyed a continuity of application by the long line of social critics in the literary tradition, but in a way that had eventually depoliticized the analytic perspective it had for so long grounded. On the other hand, the delegitimation of both of the major contemporary approaches to social analysis — in the West, the quantitative, comparative method of the Columbia School and from the East, the orthodox base-superstructure model — opened a space for the deployment of 'culture', repoliticized through efforts of the first New Left, as a primary analytic category.

The geopolitics of the mid-1950's came together with the 150-year trajectory of British (social) criticism (e.g., Burke and Paine, Arnold, Pater, Leavis) in the formation of the first New Left and the emergence of cultural studies. Raymond Williams, E. P. Thompson, and Richard Hoggart, each in his own way, contributed to the intellectual and scholarly (as well as political) bases of the movement(s). Notwithstanding the differences they exhibited in practical politics or analytical emphasis, the national perspective and common concerns for education, popular cultural forms and the 'lived' experience of class, which they shared, represent a continuity in English cultural criticism that survived in the immediate directions cultural studies took.

In Hoggart, Williams, and Thompson, the elite, literary/critical tradition, drained of politics, intersected and combined with the theoretical and practical commitment to both politics and history at the popular level on the Left. A renewed and refocused interest in the working class appeared as a common theme in the literature associated with the first New Left and claimed as formative by cultural studies: *Universities and Left Review* greeted Richard Hoggart's *Uses of Literacy* with four articles in a special section of its second issue in 1957.[33] In the tradition of Leavis and Scrutiny and in reference to the 'culture debate' — a continuity with a century and a half of humanistic (and conservative) social critique — Hoggart's recuperation of values and meanings contingent on the rejection, or upending, of the high/low culture distinction by reading working-class, popular, culture as a 'text', was decisively innovative and informed first New Left discussions of both the base-superstructure and class and classlessness questions in the 1950's.

From 'below', Hoggart recovered an urban working-class experience that was being undermined by commercialism and validated it in the face of elite cultural expressions. From 'above', Williams, in *Culture and Society: 1780–1950*,[34] exposed how the forces of reaction had appropriated texts of resistance. Both, however, were limited by the gradual exclusion of the politics of struggle from the tradition of criticism informing their work, a point often made by Thompson. Williams and Thompson shared a dedication to historical analysis inherent to Marxism, but neither Williams nor Hoggart seemed to be writing just history, or sociology, or literary criticism for that matter. Williams and Hoggart approached their subject matter from the humanities, but their work rejected disciplinary boundaries, and the disciplines rejected them. All the same, their work, with that of Thompson, collectively legitimated a return to class politics.

During the mid-1960's, it was in the academic context that the alliance developed between the conservative tradition of social critique and literary-critical methods and the first New Left concern for a positive, and political, re-evaluation of popular culture and individual agency. In different ways, both were 'humanist' and both emphasized 'values'. The early years of the institutionalization of cultural studies in the Centre for Contemporary Cultural Studies at the University of Birmingham (founded by Hoggart in 1964) extended the program begun in *The Uses of Literacy*. That is, popular culture, or working-class culture assailed by mass culture, was distinguished from elite or middle-class culture(s) and read through the methodological lenses of the literary-critical tradition to ascertain the field of values in play.

Hoggart divided the new field into three parts, drawing on both the humanities and the social sciences: these were the historical and philosophical, the sociological, and, most important, the literary critical.[35] For Stuart Hall, cultural studies was a "conjunctural practice [...] developed from a different matrix of interdisciplinary studies and disciplines" and "emerged precisely from a *crisis* in the humanities", the extension of Leavis's project of taking questions of culture seriously, by Hoggart and Williams. In fact, interdisciplinarity at CCCS did not mean "a kind of coalition of colleagues from different departments" but rather a decentering or destabilization of "a series of interdisciplinary fields" (Ref. 36, p. 11, 12, 16). The recognition of the inadequacies of received categories of analysis, the emphasis on the relationality of the field, and this decentering and destabilization of the naturalized, taken-for-granted separation of the humanities and the social sciences and the divisions among the social sciences has been fundamental to the cultural studies project with its emphasis on values and interpretation in social analysis.

From the social sciences, sciences studies (including 'sociology of scientific knowledge', 'social studies of science', and 'science and technology studies') has offered 'exogenous' appraisals of the development of science ranging from the way the social field influenced the directions of the development of science to the contingency of scientific knowledge, its (social) constructedness, and its local situatedness (e.g., Ref. 37). New disciplinary/departmental groupings have also been invented, challenging the fact-values divide and illustrating how essentialist categories of difference have functioned to subordinate entire groups. Diverse strands of 'feminism and gender studies' have pointed to the constructedness of the categories of 'man' and 'woman' and have noted how scientific discourse of the female body has functioned to position women in society. Many scholars of 'race and ethnicity' in the West have attacked essentialism as well while disputing Western universalism and objectivity. By the same token, many of those studying 'non-Western societies' have emphasized alternatives to the Western development model and pointed to the implications of these alternatives for epistemology. The 'culture wars' and the 'science wars' are striking evidence of the depth of the resistance these new developments have encountered (see Refs. 38–39). They are evidence of the fundamental nature of the debates and conflicts in the modern world over how valid knowledge may be produced, what are the grounds and the scope of its authority, and who may speak in its voice. Most importantly, the developments across the superdisciplines raise the concrete question of what contemporary social action may be considered legitimate.

The emphasis in complexity studies — on contingency, context-dependency, multiple, overlapping temporal and spatial frameworks, and deterministic but unpredictable systems displaying an arrow-of-time — suggests, as some scientists are beginning to say, that the natural world as they now see it is beginning to look unstable, complicated, and self-organizing, a world whose present is rooted in its past but whose development is unpredictable and cannot be reversed. In short, it is beginning to more closely resemble the social world, rather than *vice versa*. The question remains, if complex behavior is not amenable to explanation through hypothesis testing and theory construction because such systems, including now social systems, albeit deterministic, are inherently unpredictable, how can we proceed? Indeed, the hypothesis testing mentioned earlier may be part of the contemporary epistemological quandary. It is part of a framework oriented towards the development of explanations for particular phenomena and the generalizations on which such explanations rest. This framework includes the corollary that once the theory is known, outcomes of specific interventions can be predicted and, therefore, social science knowledge can be exploited in policy-making. This was the basis for the cross-national comparative research (the analogue of laboratory studies in the natural sciences) on which modernization theory depended. By the 1960's, the empirical failure of this work called seriously into question its theoretical and methodological underpinnings (not incidentally, just as the world-economy entered a period of contraction, U.S. hegemony came to an end, and the politics of knowledge formation became an issue for antisystemic movements).

One answer to how to proceed, at least as proposed for natural systems, is simulation. Simulations allow for the exploration of the parameters of possible, and thus also impossible, future system evolution or behavior. But what of knowledge of social reality and the social sciences? Although there is some work being done on simulation in the human world, the larger point remains that of individuating possible futures, or what Immanuel Wallerstein has called 'utopistics'[40] and what Ilya Prigogine lamented as his fear 'of the lack of utopias'[41]. Embodying an analogy between narrative and simulation, social analysts may henceforth feel licensed by the developments in complexity studies to make the shift from fabricating and verifying theories to imagining and evaluating the multiple possible consequences of diverse interpretative accounts of human reality and the actions they entail. Herein lies an alternative for a unified historical social science to the predictive, Newtonian model of social scientific inquiry. It constitutes a mode of constructing authoritative knowledge of the human world, which is of engaging in science, by producing defensible accounts and future scenarios, without chasing the chimera of predictability.

Contemporary events in our globally integrated world have shown that methods that specify (often only implicitly) an exemplar and then endeavor to predict the impact of interventions designed to move supposedly autonomous units towards some ideal state perform poorly. This is what both scholars and policy-oriented analysts are experiencing, again today, to their dismay. All the same, large-scale regularities do persist over time and particularistic 'rich description', or interpretive accounts based on an understanding, *verstehen*, of local value contexts, or resorting to 'human creativity' or 'free will' explanations fail as well to capture the interrelatedness of structure and emergence. Furthermore, an assessment of systems approaches uncovers an inherent paradox of social analysis: "the accumulation of knowledge often leads to a utilization of that knowledge both by the social scientists and the objects of their research — which may change the validity of that knowledge" (Ref. 18, p. 19). One reading of this is that social knowledge is not universal, but knowledge for specific times, today for our times, fitting nicely with findings in complex systems research and the realization that human interactions constitute such a complex, deterministic but unpredictable, system. Necessity and chance can no longer be viewed as mutually exclusive options in social research.

The combination of the conviction that there is a 'real' world and that the future, although it is 'determined' by the past, is nonetheless unpredictable and the parallel assaults on dualism (e.g., Refs. 31, 42) challenge the epistemological status of the sciences as unique discoverers, guardians, and purveyors of valid knowledge, that is, truth, by redefining what it means to describe the evolution of natural systems. As Ilya Prigogine has argued, the "sciences are not the reflection of a static rationality to be resisted or submitted to; they are furthering understanding in the same way as are human activities taken as a whole" (Ref. 43, p. 3). He goes so far as to state that "I believe that what we do today depends on our image of the future, rather than the future depending on what we do today" (Prigogine in Ref. 44, p. 28).

The fundamental importance of the crisis in knowledge formation is underlined when we must recognize that what has been going on in the sciences had been happening in analogous movements across the entire range of the structures of knowledge since the 1960's. The equilibrium, consensus model that, it was argued, wrote history and power out of the equation, was contested by intellectuals and activists alike. The theorizations of the challenges to the liberal order (such as the Vietnam War, and the civil rights, feminist, and student movements) were outgrowths of and contributed to challenges to the prevailing structure of knowledge in the long term. Beginning in this period, work from across the disciplines led to conclusions and interpretations incompatible with the premises on which the relational structure of the natural sciences, the social sciences, and the humanities were founded. Developments that indicate a collapse of the frontier separating the humanities from the social sciences have included widespread methodological ecumenism, the rise of structuralism and the concomitant recognition of 'values' as an integral part of all knowledge formation, the renewal of an appreciation of the significance of the local and the complex, and the revival of an emphasis on contingency and temporality associated with agency and creativity.[32]

Across the disciplines these arguments may be represented as a concern for spatial-temporal wholes constituted of relational structures representing the persisting regularities normally associated with a 'scientific' approach on one hand and, on the other hand, the phenomenological time of their reproduction and change (the ineluctable reality of the arrow-of-time) that captures the play of incommensurable differences associated with a 'humanistic' approach. Difference, of course, involves values. We are thus presented with a re-fusing of 'is' (the realm of facts and the goal of science) and 'ought' (the field of values and the challenge of the humanities) in the construction of systematic knowledge of human reality.

Values no longer need be, must no longer be, construed simply as a matter of individual ethics or morality in the creation of authoritative knowledge in the social sphere, but must hereafter be conceived as an integral part of a historical social science. Indeed, authoritative knowledge thus constructed would have no pretensions of universality (validity for all times and places) but rather offer defensible interpretations for particular times and places. Thus, a social science for our times, of necessity singular and transcending disciplinary boundaries, must do two things. First, it must be premised on the indissoluble unity of the regularities of social relations, their structure, and change, their history. Secondly, it must recognize that the latter supposes the integration of values as integral to inquiry, not simply as a matter of the personal inclination of the analyst.

In this secular crisis of the structures of knowledge, it is clear that the grand intellectual antinomies that have been the subject of hot debate for so long — holism versus reductionism, structure versus agency, determinism versus freedom, order versus disorder, fact versus value — are dependent not just on contradictory epistemological positions, but more surely on a specific ontological view of the world as made up of fundamentally deterministic and predictable systems. We are not living the 'new world order' but a transition period of 'new world disorder', a time of massive fluctuations far from equilibrium in the language of complexity studies. Change will not depend only on our normatively motivated action for its initiation. During a period of wide fluctuations in the constitutive processes of a system driven far from equilibrium, including a system of social relations, small fluctuations can have enormous impact even to the extent of effecting total systemic transformation — instabilities expanding possibilities, that is, opportunities, by reducing constraints. By the same token, the direction of change will, as complexity studies show, be exquisitely dependent on small fluctuations, for instance, in the form of our value-laden decisions and actions.[32,45] This is the time to *make the future* — precisely because everything is in flux. This is the time that Immanuel Wallerstein calls transformational timespace, that of *Kairos*, that of human choice "when free will is possible" (Ref. 46, p. 147). It is not so much the simple return of agency, but the manifestation of the fundamental relationship between agency and structure — the indivisibility of chance and necessity.

As I have argued elsewhere,[47] the definition of valid knowledge claims in terms of 'who, what, when, where, why' and the 'view from nowhere' is giving way. We may (indeed, I would argue we must), without forsaking our dedication to 'science', or even *technē*, turn our attention to producing authoritative knowledge in terms of 'for whom, for what, for when, for where' and 'from whose point-of-view'. The consequences of developments across the disciplines impinge directly on the manner in which scholars make claims for the legitimacy of their interpretations of social reality, and thus amount to overturning the dominant model, of knowledge and reality, shaping our understanding of the human world.

There is no way, however, that we can know if the transformation underway at present, and in which we will all play our part in one way or another, will result in a more substantively rational human world. A knowledge society we have been, and will remain. But the structure of that knowledge is already changing with understandings of the natural world and human reality evolving on non-contradictory bases. Today, in a world at the end of certainly, choice matters: conscious, cognizant choice in variously assembling our exploding stock of information to create knowledge that is both historical and systemic in the form of possible alternative futures along with their inherent value orientations. The issue then truly is choice, choice among those alternatives, the necessity of exercising it, how to exercise it, and on what basis to exercise it to bring about that more substantively rational world.

References

1. R. E. Lee and I. Wallerstein (Coords.), *Overcoming the Two Cultures: The Sciences versus the Humanities in the Modern World-System*. Boulder, CO: Paradigm (2004).
2. M. Poovey, *A History of the Modern Fact: Problems of Knowledge in the Sciences of Wealth and Society*. Chicago: Univ. of Chicago Press (1998).
3. R. E. Lee, The 'Third' Arena: Trends and Logistics in the Geoculture of the Modern World-System. In: W. Dunaway (Ed.), *Emerging Issues in the 21st Century World-System*. Westport: Greenwood, 120–127 (2003).
4. J. S. Mill, *The Logic of the Moral Sciences*. La Salle, IL: Open Court, 1988 [1872] (1843).
5. T. H. Huxley, Science and Culture. In: *Science and Education. Vol. 3 of Collected Essays*. New York: Greenwood Press, 1968, 134–159 (1881).
6. F. Machlup, *The Production and Distribution of Knowledge in the United States*. Princeton, NJ: Princeton University Press (1962).
7. D. Neef, *The Knowledge Economy*. Boston: Butterworth–Heinemann (1998).
8. P. Cooke, *Knowledge Economies: Clusters, Learning and Cooperative Advantage*. London: Routledge (2002).
9. Z. J. Acs, H. L. F. de Groot and P. Nijkamp (Eds.), *The Emergence of the Knowledge Economy: A Regional Perspective*. Berlin: Springer-Verlag (2002).
10. L. Thurow, *The Future of Capitalism: How Today's Economic Forces Shape Tomorrow's World*. New York: Penguin (1996).
11. P. F. Drucker, *Post-Capitalist Society*. New York: HarperBusiness (1993).
12. T. K. Hopkins, I. Wallerstein et al., *The Age of Transition: Trajectory of the World-System, 1945–2025*. London: Zed (1996).
13. W. R. Ashby, General Systems Theory as a New Discipline. In G. J. Klir (Ed.), *Facets of Systems Science*. New York: Plenum, 249–257 (orig. 1958) (1991).
14. W. Weaver, Science and Complexity. *American Scientist* **36** (4), July–Aug., 536–544 (1948).

15. L. von Bertalanffy, *General System Theory: Foundations, Development, Applications*. New York: George Braziller (Revised Edition, 1968).

16. N. Wiener, *Cybernetics: or Control and Communication in the Animal and the Machine. Second Edition.* Cambridge, MA: MIT Press, 1961 (1948).

17. A. Rapoport, The Organismic Direction in General System Theory. Unpubl. paper delivered at the 5th European School of Systems Science, Centre interfacultaire d'études systémiques, Université de Neuchâtel, Switzerland, Sept. 7–11 (1998).

18. F. Geyer, The Challenge of Sociocybernetics. *Kybernetes* **24** (4), Apr., 6–32 (1995).

19. S. Aida et al. (Eds.), *The Science and Praxis of Complexity: Contributions to the Symposium Held at Montpellier, France, 9–11 May, 1984.* Tokyo: United Nations Univ (1985).

20. H. R. Pagels, *The Dreams of Reason: The Computer and the Rise of the Sciences of Complexity.* New York: Simon and Schuster (1988).

21. D. L. Stein (Ed.), *Lectures in the Sciences of Complexity: The Proceedings of the 1988 Complex Systems Summer School Held June–July, 1988 in Santa Fe, New Mexico.* Redwood City, CA: Addison-Wesley (1989).

22. R. E. Lee, Readings in the 'New Science': A Selective, Annotated Bibliography. *Review* **15** (1), Winter, 113–71 (1992).

23. M. M. Waldrop, *Complexity: The Emerging Science at the edge of Order and Chaos.* New York: Touchstone/Simon and Schuster (1992).

24. D. Byrne, *Complexity Theory and the Social Sciences: An Introduction.* London: Routledge (1998).

25. J. Casti, *Complexification: Explaining a Paradoxical World through the Science of Surprise.* New York: HarperCollins (1994).

26. Sir J. Lighthill, The Recently Recognized Failure of Predictability in Newtonian Dynamics. *Proceedings of the Royal Society of London* **407** (1832, Sept. 8), 35–48 (1986).

27. I. Ekeland, What is Chaos Theory? *Review* **21** (2), 137–150 (1998).

28. G. Nicolis and I. Prigogine, *Exploring Complexity: An Introduction.* New York: W. H. Freeman and Co (1989).

29. I. Prigogine and I. Stengers, *Order out of Chaos: Man's New Dialogue with Nature.* New York: Bantam (1984).

30. I. Prigogine and D. J. Driebe, Time, Chaos, and the Laws of Nature. In: E. R. Infeld, R. Zelazny and A. Galkowski (Eds.), *Nonlinear Dynamics, Chaotic, and Complex Systems.* New York: Cambridge Univ. Press, 206–223 (1997).

31. I. Prigogine, *The End of Certainty: Time, Chaos, and the New Laws of Nature.* New York: Free Press (1996).

32. R. E. Lee, *Life and Times of Cultural Studies: The Politics and Transformation of the Structures of Knowledge.* Durham, NC: Duke Univ. Press (2003).

33. R. Hoggart, *The Uses of Literacy: Aspects of Working-class Life, with Special References to Publications and Entertainments.* London: Chatto and Windus, 1967 (1957).

34. R. Williams, *Culture and Society: 1780–1950*. New York: Columbia University Press, 1983 (1958).

35. R. Hoggart, Schools of English and Contemporary Society. In: *About Literature. Vol 2 of Speaking to Each Other: Essays by Richard Hoggart*. New York: Oxford University Press, 1970 (1963).

36. S. Hall, The Emergence of Cultural Studies and the Crisis of the Humanities. *October* **53**, Summer, 11–23 (1990).

37. S. Shapin, Here and Everywhere: Sociology of Scientific Knowledge. *Annual Review of Sociology* **21**, 289–321 (1995).

38. R. E. Lee, The 'Culture Wars' and the 'Science Wars'. In: R. E. Lee and I. Wallerstein (Coords.), *Overcoming the Two Cultures: The Sciences versus the Humanities in the Modern World-System*. Boulder, CO: Paradigm, 189–202 (2004).

39. Boaventura de Sousa Santos (Ed.), *Conhecimento Prudente para uma Vida Decente: Um Discourso sobre as Ciencas Revisitado*. Porto: Afrontamento (2003).

40. I. Wallerstein, *Utopistics: Or, Historical Choices of the Twenty-first Century*. New York: New Press (1998).

41. I. Prigogine, Beyond Being and Becoming. *New Perspectives Quarterly* **21** (4), Fall, n.p. (2004).

42. J. D. Barrow, *The Artful Universe: The Cosmic Source of Human Creativity*. Boston: Back Bay (1995).

43. I. Prigogine, The New Convergence of Science and Culture. *UNESCO Courier*, May, 9–13 (1988).

44. M. B. Snell and Y. Yevtushenko, Beyond Being and Becoming. *New Perspectives Quarterly* **9** (2), Spr., 22–28 (1992).

45. R. E. Lee, After History? The Last Frontier of Historical Capitalism. *Protosoziologie* **15**, 87–104 (2001).

46. I. Wallerstein, *Unthinking Social Science: The Limits of Nineteenth-Century Paradigms*. Cambridge, UK: Polity (1991).

47. R. E. Lee, Sesión Inaugural. Inaugural Address, Third International Conference on Sociocybernetics, Spanish–English Parallel Text. *Journal of Sociocybernetics* **II** (2), Fall/Win., 43–48 (2001/02).

ON BRIDGING THEORY AND PRACTICE
IN THE PERSPECTIVE OF HISTORY

ELLEN VAN KEER

Leo Apostel Centre for Interdisciplinary Studies
Vrije Universiteit Brussel (VUB), Belgium
E-mail: evankeer@vub.ac.be

The gap existing between theory and practice is an old sore and a fashionable criticism of many disciplines in arts and humanities. The historical studies are no exception in this. The traditional approach is evidence-based and concentrates on practical and empirical matters while abandoning problems of interpretation and theory. It leaves questions of epistemology to the branches of theory and philosophy of history and science. However, the rise of 'postmodernism' brought to light that non-empirical factors are inevitable and pose major epistemological obstacles in the perspective of history. In particular, it has become clear that historians cannot but translate the past into their own words and concepts, and must be critic of hidden linguistic and cultural factors actively shaping the knowledge they produce. Indeed, a disciplined disregard of historical theory in historical writing is rooted in disruptive assumption underlying the notion of 'theory' in itself. To deal with the problem, this essay proposes a critical examination of the role and concepts of (1) epistemology, (2) theory, (3) philosophy, (4) methodology, and (5) historiography in the study of history. Historiography emerges as a most promising subject and field to expose and address issues of theory in the practice of writing history. Furthermore, (6) archaeology traditionally describes the empirical-analytical study of past material objects but recently it was also developed into a critical-theoretical concept that exposes hidden conceptual foundations, implicit modern assumptions, and intrinsic linguistic structures determining our knowledge about the past. Thus, it integrates the main opposing and complementary tenets and approaches in the 'new' perspective of history.

1. Introduction

The gap existing between theory (*episteme*) and practice (*techne*) is an old sore and fashionable criticism of many disciplines in the arts and the humanities. The historical studies are no exception in this. Art historians, for instance, are mainly engaged in the practice of collecting, describing and classifying objects of art. While focussing primarily on practical and empirical matters, they largely pass by problems of interpretation and theory. This leaves issues of epistemology to the branches of theory and philosophy of history and science. However, it recently emerged that theory and practice cannot entirely be separated and practicing historians must be aware and critic also of non-empirical factors shaping the knowledge they produce. In this essay, we will first show how the disregard of (1) epistemology in historical writing is rooted in assumptions underlying the notions of (2) theory and (3) philosophy. Next, we will explore how the subjects and fields of (4) methodology and (5) historiography relate to theory and practice in the writing of history. Finally, we will show how the notion of (6) archaeology can refer both to empirical analysis and to epistemological criticism. Our central argument is both are equally fundamental in the perspective of history.

2. Epistemology

2.1. *General Epistemology*

Epistemology is a compound term built on the Greek words *episteme* i.e., 'knowledge' and *logos* in the (modern) meaning of "reason, science, study".[a] Its most simple and general definition is 'study of knowledge'. In scholarship, it primarily refers to a branch of philosophy that concentrates on problems and distinctions in relation with the nature and extent of human knowledge. It involves questions such as: "Is knowledge a-priori or a-posteriori?", "Is knowledge based on deductive or inductive forms of reasoning?", "Is (the capacity for) knowledge innate or acquired?", "Is knowledge purely cognitive or is it also behavioural?", "Is knowledge universal and absolute or temporal and relative?", "Is knowledge identical with Justified True Belief?", etc. Epistemology has both descriptive and normative functions. As of old, rationalism (as idealism) and empiricism represent the main opposing tenets. In short, the former defends that all knowledge lies in conscious and reason while the latter defends that all knowledge originates in experience and the senses. Recently, constructivism offered a tempting alternative to polarizing the issue into objectivism against subjectivism. It argues that reality is neither purely mental nor strictly physical but 'mediated' through perception and sign systems by which we experience and express it. There is no aim in searching the ultimate single true meaning or essence of things that is permanent and universal. Meaning and knowledge

---

[a]Originally *logos* had a wider meaning associated with *legein* i.e., "to speak, to say". It acquired its more narrow meaning of 'science' with the emergency of scientific discourse to which the development and concept is constitutive.[1]

are human dependent and culturally determined and need to be located within their specific contexts or origination and signification. The position encouraged developments such as 'epistemology of perception' and 'social epistemology', which observe respectively the role of human perception and social contexts in processes of knowledge acquisition. For some this represents a radical departure from classical epistemology while for others it merely enlarges its scope. A clear consensus on the definition of the field remains unavailable.

The focus and position of epistemology are also subject to debate and change. Classical epistemology concentrated on the pursuit of truth and on the question of "what is (true) knowledge?" It closely related to ontology, the branch of philosophy that specialised in questions of "what is?" However, in the Anglo-American tradition, for instance, we find that interest in ontology (and metaphysics generally) has declined while the focus of epistemology has shifted away from the problem of 'knowing that' i.e., 'factual' knowledge towards the problem of 'knowing how' i.e., 'pragmatic-contextual' knowledge. It raises questions such as: "How do we acquire and produce valid scientific knowledge?", "What methods can/need we use?", "How do we assess the validity of scientific theories and interpretations?", etc. Interest in the foundations and the limitations of producing scientific knowledge is growing fast and observable notably in the successful new branch of 'philosophy of science' which investigates the metaphysical, epistemic and semantic aspects of science. Epistemology as 'philosophy of knowledge' or 'theory of knowledge' has a key role in 'philosophy of science' and 'theory of science' as well as it enjoys unprecedented individual success in this form.[2]

2.2. *History and Epistemology*

Outside the disciplines of philosophy (of science), fundamental epistemological questions are raised in new domains such as quantum physics or neuropsychology, which go beyond the traditional limits of science and scientific knowledge. However, in most disciplines epistemology assumes the form of 'theory of knowledge' applied to the particular subject and field. Likewise, philosophers separate general epistemology from epistemology (and philosophy) of mathematics, physics, psychology, art, religion, etc. Epistemological questions that apply specifically to history, the main concern here, are: "Is the past a knowable and representational reality?", "What is the historical truth?", "Is historical knowledge permanent?", "Can historians — as subjects — achieve objective or true knowledge about the past?", "How do historians produce adequate accounts and valid explanations of past events?", "What premises, approaches, methods and sources need/can they use?", "Which are valid criteria to evaluate the conclusions?", etc. Historians' engagement in the theory of historical knowledge has rapidly expanded in recent years because the postmodern context has posed fundamental challenges to the scientific foundations of traditional historical writing and to the possibility of objectivity especially. The view of classical 'historicism' puts forward that empirical analysis of historical sources ('evidence') results in objective ('true') knowledge about historical reality ('facts' and 'events'). Its core values are often summarised by the catchphrase of "how it really was". This expression derives from a statement made by the 19th century German scholar Leopold von Ranke (1795–1886), after whom we also speak of the 'Rankean' paradigm in history.[b] However, new ideas have emerged that are in radical disagreement with the underlying credo of realism. To the contrary, it is now argued that we have no objective access to reality because linguistic and cultural barriers are inevitable and subjective. The new argument is associated with 'postmodern' theories that exposed the fundamentally narrative and subjective (i.e., 'discursive') nature of our knowledge about the world and every thing in it. In this view, there is no such thing possible as objectively knowing and representing the past because historical accounts and even the historical 'facts' on which they build inevitably involve interpretation and this introduces subjectivity. In extremis, the validity of historical writing and the possibility of historical knowledge are totally denied on the grounds that the dichotomies of fact/fiction and true/false do not hold and the past is essentially an 'invention' or a 'construction' of the present.

As expected, these claims and developments engendered a truly enormous and highly contentious literature, causing 'postmodernism' to become extremely difficult to define, even within the limited framework of the historical studies. For Anglo-American historians, who are the focus here, 'postmodernism' refers in a most narrow and controversial sense to the ideas associated with postmodern 'critical' theory and with continental 'poststructuralist' philosophers such as Michel Foucault (1926–1984) and

---

[b] "You have reckoned that history ought to judge the past and to instruct the contemporary world as to the future. The present attempt does not yield to that high office. It will merely tell how it really was."

Jacques Derrida (1930–2004) more particularly.[c] Their intellectual attacks against realism and objectivism announce the 'end of history' in the classical empiricist tradition, where they are instinctively negatively received and on occasion countered with equal radical anti-criticisms.[d] However, in a broader and more optimistic sense, 'postmodernism' is associated with the rise of a new scholarly mentality that is characterized by growing epistemological concerns and an increasing awareness of the linguistic and cultural determinedness of reality and meaning more particularly. It assumed the name of the 'linguistic turn' in the humanities.[e] Historians too have become increasingly aware of the problem and postmodern historical theory is on a steep rise.[f] However, most historians tend not to take it so far. In practice, we sense a new mentality mostly in rapidly increasing varieties of sources, methods, subjects, approaches and perspectives adopted in historical writing and most markedly in the so-called 'New History' and 'new (cultural) histories'.[g] On these developments, cf. Ref. 32. Conversely, debates and reflections on the nature and purpose of history and on the scientific foundations and limitations of historical knowledge tend to remain confined to a relatively isolated branch that is concentrated specifically on theory and philosophy in history and attended to only by a limited group of historians.[33] Practicing historians sometimes call this the realm of history

---

[c]However neither thought of themselves as 'postmodernists' or 'critical' theorists. Moreover, 'critical theory' is a generic term for theories critic of culture and knowledge, a feature associated notably with postmodernism for which it became in turn a generic term. 'Critical' theories came originally from a variety of fields, including continental 'poststructuralist' philosophy, social and cultural theory, or literary criticism. They diverge on many points. Characteristically, however, 'postmodern' critical theory exposes the relativist and subjectivist effect of language and culture in science and claim that (a) 'reality' is unrepresentable and (b) 'truth' unachievable.[3] For implications and applications in history, cf. Refs. 4–7.

[d]E.g., Refs. 8, 9. More generally, cf. Ref. 10.

[e]This expression was given currency by the American philosopher (of language) Richard Rorty (1931–2007).[11]

[f]Off the continent, it kicked in with the foundation of the journals *Past and Present* (1952) and *History and Theory* (1960). For core publications, cf. Refs. 12–21.

[g]The notion of *Nouvelle Histoire* derives from the French so-called *Annales* school and its third generation of scholars more particular. Starting in the '30ies, this movement opened up traditional writing of (biographical, political, military) history to new subjects, problems, methods, and approaches derived from the discipline of sociology (and anthropology). Later disappointment with social history and theory (Marxism especially) opened the door for ideas derived from 'critical' cultural and literary theory. This lead up to the so-called 'cultural turn' in historical writing off the continent in the '80ies, where it is most markedly sensed in the rise of the new 'cultural' history or histories. For core publications, cf. Refs. 22–31.

186

in the 'second' sense because it seems more concerned with problems related to the contemporary study of the past rather than with actually studying the past. They set it against historical writing as focused on history in the 'first' sense and contending to the empirical study of the past 'in its own sake'. Others distinguish similarly between history 'e' and history 'a' or history as past 'event' and present 'account' or differentiate between history with small 'h' as referring to the past and with capital 'H' as describing the study of the past. All these distinctions, however, presuppose that the past exists entirely independent from us and they suggest furthermore that we can provide a totally objective account of it. We realise now this is not the case.

3. History and Theory

Even practicing historians use theory and interpret all the time. Every model that expresses and structures correlations between observations and reflections, every pattern of significance found in historical developments, every contemporary explanation etc. is essentially a theory or an interpretation. To conceive of 'prehistory' (already by adopting this term) as a linear progressive evolution from Stone Age over Bronze Age to Iron Age in association with growing cognitive capacities and efficiency of control man gained over his environment, is one vision on the course of history (i.e., evolutionism). Another view explains historical developments as the result of contact and influence between different species and cultures (i.e., diffusionism). Recent models have abandoned the idea of simple causal relations (i.e., teleology) in favour of more complex processes of interaction that are the result of many and multiple natural and cultural factors (i.e., processualism) and about which science is a current form of communication (i.e., systems theory).[h] Likewise, seemingly innocent concepts and words historians randomly use, e.g., freedom, religion, marriage, world, music, jewellery, pig, etc., will carry inevitable anachronistic assumptions and affect their perception and interpretations of the past. As a result, no historical explanation or description will ever be completely theory-free and value-free. However, many historians fail to appreciate the contingent and conceptual element inherent in historical writing. In fact, for most practicing historians, the writing of history has preferably nothing to do with theory or interpretation at all.

---

[h]We currently notice a growing concern for theory also in archaeology and it equally tends to take the shape of a separate branch or specialization. For core publications, cf. Refs. 34, 35. More generally, cf. Ref. 36.

Diverging meanings and assumptions attached to the notion of 'theory' significantly complicate the issue and make this a perfectly good example of the nature and impact of hidden linguistic difficulties addressed in historical theory. From the Greek word *theoria* we derived the word 'theory' in preserving the meaning of 'supposition'. In common language, we use it to describe four things. (1) The sum of ground rules and abstract principles of a certain art, technique or science (e.g., the theory of music), (2) A system of ideas or hypotheses accepted to explain something (e.g., the theory of evolution), (3) Abstract thought and ideal belief taking no account of facts or feasibility (e.g., "In theory every university should offer a course on ancient Greek music"), (4) Unproved assumption and conjectural guessing (e.g., "The theory Saddam Hussein produced nuclear weapons"). We find parallel usages in the historical sciences. Only here it appears that a majority of historians oppose to theory because they primarily understand this as referring to unfounded suppositions (speculative = 4) opposed to reality (ideal = 3) and even more specifically to Universalist models of explanation in contrast with the perceptible arbitrariness of the empirical past (conceptual = 2). In its negative acceptation, the term is particularly associated with 'metaphysical' theories produced disciplines such as philosophy (e.g., Hegel), sociology (e.g., Marx), biology (e.g., Darwin), psychology (e.g., Freud), anthropology (e.g Lévi-Strauss), linguistics (e.g., De Saussure) etc. Conversely, the historians who endorse theory appear to understand it primarily as referring to the premises and principles of historical inquiry (technical = 1). In its positive reading, the term relates to the theory of historical knowledge and appeals furthermore to assess the foundations and limitations of historical knowledge, the impact of concepts and their definitions, the workings of interpretation and their implications, etc. — viz. to respond to the challenges posed by the postmodern 'critical' movement. Some set this as theory with small 't' against theory with capital 'T'. Most historians are thinking of 'Theory' and regard it as nonessential or even opposed to the empirical practice of historical writing. However, historians thinking of 'theory' maintain that it is inevitable and a mandatory concern in any historical inquiry. Similarly, the connotations of 'subjective' attached to the term 'interpretation' have encouraged its suppression from historical writing while making it a central concern in 'theoretical history'.[i]

---

[i]Ref. 37, s.v. 'Theory', pp. 347–349, Ref. 38, s.v. 'Interpretation', pp. 243–250.

4. History and Philosophy

The controversy over theory has historical and ideological grounds complicating also the attitudes towards the related subject and field of philosophy in history. Broadly speaking, we can parallel the 'metaphysical' and 'critical' implementations of 'theory' with contrasting 'idealist' and 'analytic' orientations in 'philosophy'. Historians generally stress the former, for being the oldest and incorporating an approach to which the historical sciences fundamentally oppose. The scientific study of history was established in the empiricist tradition of the 19th century as focused on inductive analysis of particular past events and facts, in opposition and reaction to the hypothetic-deductive approach prevailing in (German) Idealist philosophy at the time. This centred on speculative metaphysics which aims to reveal the overall meaning and general principles of the totality of all time and being.[j] 'Philosophy of history' classically refers to speculative philosophies of history that endorse the historical process as a whole and seek to explain its entire being and meaning (e.g., Hegel). Obviously, this approach contrasts sharply with the perceptible arbitrariness of the empirical reality endorsed in the historical sciences. Moreover, taken together, speculative theories produce a fundamental paradox. On the one hand, they all imply and complement each other, yet on the other hand, each pretends to universality in itself. Karl Popper (1902–1997), an Austrian-British 'analytic' philosopher, achieved everlasting fame for having established the fundamentally hypothetic and unverifiable nature of scientific theories from this.[39] Nevertheless, Popper also rejected the possibility of radical empiricism endorsed in the scientific study of history. Science, he argued, is *tout-court* to a certain degree hypothetic and 'unverifiable' i.e., not able to prove right. It is, at its best 'falsifiable' i.e., able to prove wrong, in as many aspects as possible. Popper's now widely accepted 'falsifiability' criterion for scientific theories and, by extension, all the theoretical elements inherent in any scientific inquiry, states that they must not simply pass experimental and empirical tests but allow for maximum 'falsification' by stating clearly and precisely the premises and principles underlying them.[40] 'Analytic' philosophy shifts the focus away from producing essentialist scientific explanations towards analyzing the foundations and limitations of scientific concepts and knowledge. Accordingly, 'analytic' philosophy of history investigates the pro-

---

[j]A statement of Leopold Von Ranke is again the catchphrase here: "From the particular, one can carefully and boldly move up to the general; from general theories, there is no way of looking at the particular."

cesses and structures of historical knowledge, the qualities and character-istics of historical explanations, the conditions and limitations of historical writing, the validity of historical methods and categories, etc. In addition, the 'analytic' tradition, which is the predominant Anglo-American orien-tation, has now fruitfully intersected with continental 'postmodern' theory to produce the so-called 'new' philosophy of history. Characteristically, it considers history as a form of 'discourse' and concentrates on the impact of philosophical, cultural, linguistic, and other matrices in which historical knowledge is generated.[k] Epistemological skepticism constitutes the basic feature of the merging subjects and fields of contemporary 'philosophy of history', 'historical thought', 'historical theory', and 'theoretical history'. Indeed it's one of the few more general characteristics of 'postmodernism' we find in all disciplines of the humanities.

Many historians, however, due to their old and disciplined resistance against philosophy and theory, also disincline to take op the new chal-lenges associated with critical theory and analytic philosophy. To the con-trary, the 'postmodern turn' further increased their sceptical distance to-wards the controversies over theory and philosophy because it also brought back an unprecedented and paradoxically enough not necessarily always as critical a fashion for 'Theory' and 'Philosophy' in the humanities (e.g., structuralism).[47] On the more downside, the issue has polarised in, on the one extreme, historians who adhere to often one particular 'Theory' (e.g., semiotics) and who are criticised of being postmodern intellectual-ists pampering fashionable but impracticable methodologies and escaping all accountability for this on the nihilistic grounds that no authoritative account can ever exist of anything. On the other extreme we find histori-ans who renounce to 'theory' altogether and who are criticised of becoming provincial 'empiricists' who indulge in factual detaillism in a naive quest for the objective truth. On the more positive side, however, middle ground approaches are burgeoning and an increasing number of historians are try-

---

[k]'Analytic' philosophy is a generic designation for the latest current in Anglo-American philosophy. However, rather than representing a single doctrine, there is much diver-sification in the field and there are many overlaps with other subjects and branches. A central characteristic is the rejection of convoluted philosophical systems and the emphasis on clarity and refutability of argument (i.e., Popper). Formal logic and anal-ysis of language are a central concern associated notably with logical positivism (i.e., Wittgenstein). Ordinary language is a concern associated rather with pragmatism (i.e., Rorty). For core publications of 'analytic' philosophy of history, cf. Refs. 41–43. For core publications of the 'new' history of philosophy, cf. Refs. 44, 45. More generally, cf. Ref. 46.

ing to bridge the gap between new insights derived from postmodern theory and the traditional empirical practice of writing history.[48]

5. History and Methodology

The notion and field of 'methodology' allows illustrating the issues at stake in more detail. 'Methodology' derives from the Greek expression *meta hodos* i.e., "the way by which" and *logos* i.e., "study, reason". Recently it acquired a distinct theoretical-epistemological dimension referring to conceptual descriptions of the premises, categories and methods used in historical research.[49] However, in its first and primary application it applies to the principles of 'source criticism' that represent the central scientific method of empirical-analytical historical writing.[1] In science, methodology often refers to an account of the sources, techniques and methods used for gathering information and producing conclusions in a particular scientific field or investigation. For instance, when sociologists test a hypothesis by conducting a survey, the (often introductory chapter on) methodology will account for the number of respondents considered, the parameters taken into account (sex, age, education), the kind of questions asked (yes/no, true/false, open questions, multiple choice), the type of statistics used (parametric, non-parametric), etc. Similar methods or methodologies (in this view both words are interchangeable) can be used in the historical sciences. In fact, the rise and success of 'New History' is originally to be associated with benefits gained from linking the study of history with the social sciences and from applying their quantifying methods, inductive methodologies and experimental approaches to the study of the past.

However, it remains that the central methodology in arts and humanities is not quantitative but qualitative and consists of linguistic (vs. numeric) analysis and interpretation of texts, prose, poetry, works of art, pieces of music, etc. — i.e., cultural products originating from the human mind. Objectivity is a major problem here. The arts and humanities have always struggled with an impression they give to be unscientific because they lack

---

[1]The 'historical method' was founded at the end of the 19th century.[50] It describes the techniques by which historians make use of sources and includes (a) external criticism (date, localization, authorship, authenticity, integrity, and credibility) and (b) internal criticism (historical reliability) of sources, and synthesis (explanation, inference, analogy). Basic English textbooks on the subject are Refs. 51, 52. For new academic defences of the traditional empiricist approach against the rise of new history and 'postmodernism' in history, cf. Refs. 53–60. For an overview of the entire plethora of viewpoints and debates associated with 'postmodernism' in history, cf. Refs. 61, 62.

objective sources and frames of reference. The general opinion in vigour in the scientific community is the decisive factor determining whether conclusions are acceptable or not. However, this appears to be the case to a certain degree in all sciences — even 'exact' sciences (e.g., Galileo, chaos theory). The problem becomes fundamental when the profound interpretive nature of scientific inquiry is not recognised and retaliated with in-depth interpretive criticism. This is often the case and major problem in disciplines of the arts and humanities. Philological criticism is principally text criticism. Art criticism is mainly stylistic-aesthetic criticism. Historical criticism is essentially source criticism, etc. The historical sciences rely upon solid methodologies attending to a whole range of difficulties posed by varied historical 'evidence'. With the rapidly increasing variety of sources, methods and approaches currently adopted in historical writing, the subject and field of historical methodology exploded. However, the historical sources do not speak for themselves. The historians bestow them with meaning. Nevertheless, historical methodology attends only marginally to epistemological issues and historical criticism does not usually takes into account non-empirical and contingent factors at play, leaving them to remain ill considered, unsystematic and problematic.[m]

6. History and Historiography

The notion and field of 'historiography' emerges to bridge the gap between theory and practice in the study of history more instinctively. Like history, it has a 'primary' sense (referring to a past phenomenon) and a 'secondary' meaning (describing its present study). Accordingly, it acquired empirical as well as epistemological implications. On the one end of the spectrum, historiography describes the literary genre historical writing was before the rise of the scientific study of history. 'Greek historiography' refers to the texts and study of the ancient Greek history writers. On the other end of the spectrum, historiography describes scientific historical writing. It includes three aspects defined as: (1) 'descriptive', (2) 'historical' and (3) 'analytic' or 'critical'.[n] (1) 'Descriptive' historiography studies the sources and methods used in historical writing, which is largely synonymous with the subject of historical method. (2) 'Historical' historiography studies the development of historical writing and thought, which exposes the variety and relativity of the sources, methods, concepts, premises, approaches and

---

[m]For this criticism and a more comprehensive approach, cf. Ref. 63.
[n]Ref. 64, 1–5.

perceptions historians have endorsed over time. (3) 'Analytic' or 'critical' historiography studies the scientific foundations of writing and knowing history, which is largely synonymous with contemporary 'philosophy of history' and 'theoretical history'.° Accordingly, 'postmodern historiography' describes: (1) Studies of 'new' history in practice. (2) Studies of new developments in historical writing and thought, multiplying exponentially under influence of the challenges the postmodern context has posed. (3) Studies that put principles derived from postmodern theory to practice in a critical reassessment of the foundations and limitations of knowing and writing history.P Practising science consists in developing and applying appropriate methods to gather and analyze data and in organizing the information obtained in coherent systems of signification. The academic framework we adopt and the personal assumptions we have will shape the questions we ask. The methods and sources we use will determine the answers we find. The conclusions we reach will support or demand to modify personal and general opinions, etc. Scientific knowledge is the result of a multifaceted historical process and therefore best understood in a comparative historical framework. The theory of paradigmatic revolutions developed by the American philosopher of science Thomas Kuhn (1922–1996), represents a most authoritative and ultimate assertion of the processes involved.[71] It defines a paradigm as the body of shared ideas within a specific scientific community. The historical paradigm, for instance, represents the dominant opinion of the historical academics about matters pertaining to the field of history, valid instruments for historical study, decisive factors in historical arguments, etc. Varying opinions may result in different orientations or 'schools' (e.g., social, feminist). New information, different methods or other approaches can cause a true 'revolution' and overthrow the paradigm. Classical philology, for instance, traditionally studied ancient texts in their written form but findings in the field of ethnology recently exposed their fundamentally oral nature. It shifted the focus from 'text' to 'context' and various new subjects, methods and approaches (e.g., oral history, anthropology) came into view. This produced an unseen expansion of the classical studies as well as it shaped a radical new perception of 'classical' Greece. Rather than representing the first 'literate' and 'historical' society at the beginnings of western civilisation, ancient Greece has been recognised as a

---

°For the 'descriptive' approach, e.g., Ref. 65. For the 'historical' approach, e.g., Ref. 66. For the 'analytical' or 'critical' approach, e.g., Ref. 67.
PFor (1), e.g., Ref. 68. For (2), e.g., Ref. 69. For (3) e.g., Ref. 70.

traditional 'oral' and 'ethnic' culture profoundly embedded in the ancient civilisations of the Mediterranean East. By implication, the old paradigm of 'Hellenism' has showed up as a largely modern construct (freighted with our own moral and cultural values) rather than a merely empirically recorded discovery. 'Postmodernism' deprived historical writing from the empirical facts and the objective reality that constituted its traditional scientific basis. However, in aiming to expose the temporal and contingent nature of meaning and knowledge, it bestowed the highest epistemological value upon the historical framework. In fact, the linguistic turn eventually produced a so-called 'historic turn' in the contemporary perception of knowledge and science.[72] Moreover, 'historicism', first vilified, was later reassumed as 'new' historicism in 'critical' theory and this popularised to become the dominant paradigm in the humanities.[73] Postmodern historiography is no exception as it studies historical sources no longer only in view of their material and empirical condition but also in view of their conceptual and contingent condition.[74] Ultimately, the 'new' developments mediate in themselves the disrupting and blinding dichotomies of history and philosophy, practice and theory, objectivity and subjectivity, past and present, etc.

7. History and Archaeology

In conclusion, I report that 'archaeology' has similar capacities as it endorses empirical-analytical as well as narrative-linguistic approaches to history, taking moreover both old 'realist' ideals and 'postmodern' anti-realist ideas to their limits. We all know archaeology as the scientific branch that is specialized in empirical-analytical study of material remains from past cultures, i.e., the 'realia'. Its core scientific method focuses on formal-stylistic analysis and classification of artefacts following guidelines lied down in late 19th – early 20th century handbooks and deeply marked by the empiricist-positivistic spirit of the time.[q] Moreover, the goal in archaeology is ultimately to *reconstruct* the past based on the material traces it has left and archaeologists are typically eager to realize this objective also in actual practice. Much renown of music archaeology, for instance, lies in producing playable reconstructions of ancient musical instruments and physically

---

[q]E.g., the famous handbook by the British Egyptologist Sir William Flinders Petrie.[75] Indicative of the spirit is his statement: "I believe the true line of research lies in the noting and comparison of the smallest details." By linking styles of pottery with periods, he was the first to use seriation in Egyptology, a new method for establishing the chronology of a site, which was used for dating and phasing the Aegean 'Dark Ages', 'geometric' period.

reproducing the sound of ancient music. The underlying ambition involves an extreme implementation of the values and pursuits of the 'realist' or 'reconstructionist' approach to history. For it aims to reconstruct the past in (the present) reality.[r] However, archaeology also has applications in the opposite 'anti-realist' and 'deconstructionist' approach to history. The term 'deconstructionist history' is used to describe the application in history of postmodern theories that argue against the existence of past reality independent from present perception and conceptualisations. Instead, it turns to the historical discourse in order to expose how hidden linguistic and cultural factors inform historical knowledge. The concept of 'deconstruction' derives from Derrida and involves, broadly speaking, a method of literary and cultural analysis that emphasises the internal working of sign systems, the relational quality of meaning, and the impact of assumptions implicit in forms of expression on processes of interpretation and signification.[s] Interestingly, Foucault adopted the concept of 'archaeology' for establishing the discursive nature of meaning and knowledge and breaking down intrusive assumptions and deceptive absences in science.[t] In short, he argues that language and concepts we (must) use are socially, culturally and historically determined and actively shape our knowledge and understanding of the world. He denies that meaning and knowledge can exist in extra-linguistic and extra-cultural contexts but claims to the contrary that they are linguistically, socially, culturally and historically constituted in a particular periods' *episteme*, i.e., the cultural discourses and power arrangements that regulate it. Each society has its specific regimes of truth ('politics'). Hence, the issue becomes to show 'how' (vs. 'that') these parameters constitute meaning and knowledge. Foucaults' *Archaeology of Knowledge* does precisely that.[78] It establishes the conceptual and contingent foundations and limitations of the historical discourse and relocates meaning and knowledge within their specific discursive and ideological structures. This radically shifts the focus from the content of the past to the form of history, from object to concept, from facts to interpretations, from objectivity to subjectivity, and even further, from singularity to plurality, from continuity to discontinuity, from sameness to difference, etc. Such is the linguistic-narrative perspective of history ultimately.

---

[r] On the foundations of 'reconstructionism' in history, cf. Ref. 76, pp. 194–198.

[s] Cf. Ref. 76, s.v. 'deconstructionist history', pp. 69–74.

[t] Cf. Refs. 77, 78. Note that Robin George Collingwood (1889–1943), regarded as the founder of this approach in the Anglo-American tradition, was actually engaged in archaeological research.[79]

8. Conclusion

History, as we know it, in its double meaning of the past and the present study of past, is impossible to untangle. By implication, empirical analysis and epistemological inquiry are equally fundamental to it. The perspectives, models, concepts, methods and even the facts historians use or dismiss, involve modern premises and assumptions and must be subjected to explicit scientific criticism if we ever want to disentangle and understand history in its full complexity. Historical methodology, however, is traditionally of a predominant practical-empirical nature and does not usually imply criticism of non-empirical factors. This leaves them to act as powerful disruptive forces that have a harmful blinding effect on our understanding the past. However, epistemological scepticism and methodological self-reflexivity have emerged as the central merits and characteristics of 'postmodernism' in history. The movement advantageously resulted in a critical reassessment i.e., 'rethinking' of the past, and even more fundamentally in a critically reassessment i.e., 'archaeology' of our knowledge about the past. Intrinsically, theory and practice, object and concept, past and present, fact and interpretation, history and philosophy, etc. all interrelate in intricate ways and it ultimately is these interrelations that postmodern historiography investigates.[u]

Acknowledgments

This work is supported by Fonds Wetenschappelijk Onderzoek Vlaanderen, Vrije Universiteit Brussel, and Centrum Leo Apostel. Dr. J. Wambacq (UGent) kindly enough reviewed a draft of this text.

References

1. B. Lincoln, Competing Discourses: Rethinking the Prehistory of Mythos and Logos. *Arethusa* **30** (3), 341–368 (1998).
2. J. Greco and E. Sosa (Eds.), *The Blackwell Guide to Epistemology*. Oxford: Blackwell (1999).
3. V. B. Leitch (Ed.), *The Norton Anthology of Theory and Criticism*. New York: Norton (2001).
4. K. Jenkins, *Why History? Ethics and Postmodernity*. London & New York: Routledge (1999).

---

[u]Off the continent, its success is narrowly associated with the foundation and success of *Rethinking History. The Journal of Theory and Practice* (1996), cf. Ref. 80.

5. B. Southgate, *Postmodernism in History. Fear of Freedom?* London & New York: Routledge (2003).

6. W. Thompson, *Postmodernism and History. Theory and History.* Hampshire: Macmillan (2004).

7. C. G. Brown, *Postmodernism for Historians.* London: Longman (2005).

8. B. Palmer, *Descent into Discourse. The reification of language and the writing of social history.* Philadelphia: Temple University Press (1990).

9. K. Windshuttle, *The Killing of History: How Literary Critics and Social Theorists are Murdering our Past.* New York: Free Press (1995).

10. T. Eagleton, *After Theory.* New York: Basic Books (2003).

11. R. Rorty, Metaphysical Difficulties of Linguistic Philosophy. In: R. Rorty (Ed.), *The Linguistic Turn. Essays in Philosophical Method.* Chicago: Chicago University Press, 1–40 (1968).

12. E. H. Carr, *What is History?* Harmondsworth: Penguin (1961).

13. H. White, The Burden of History. *History and Theory* **5**, 111–134 (1966).

14. L. Stone, *The Past and the Present.* Boston: Routledge & Kegan Paul (1981).

15. H. White, *The Content of the Form: Narrative Discourse ad Historical Representation.* Baltimore: Johns Hopkins University Press (1987).

16. K. Jenkins, *Re-thinking History.* London & New York: Routledge (1991).

17. K. Jenkins, *On What is History?* London & New York: Routledge (1995).

18. A. Munslow, *Deconstructing History.* London & New York: Routledge (1997).

19. B. Southgate, *Why Bother with History?* London: Longman (2000).

20. K. Jenkins, *Refiguring History: new thoughts on an old discipline.* London & New York: Routledge (2003).

21. B. Southgate, *What is History for?* London & New York: Routledge (2005).

22. L. Hunt (Ed.), *The New Cultural History.* Berkeley: University of California Press (1980).

23. P. Burke (Ed.), *New Perspectives on Historical Writing.* Cambridge: Polity Press (1991).

24. W. Melching and W. Velema, *New Trends in Cultural History.* Amsterdam: Rodopi (1994).

25. P. Burke (Ed.), *Varieties of Cultural History.* Cambridge: Polity Press (1997).

26. M. Poster (Ed.), *Cultural history and Postmodernity. Disciplinary Readings and Challenges.* New York: Columbia University Press (1997).

27. V. E. Bonnell and L. Hunt (Eds.), *Beyond the Cultural Turn. New Directions in the Study of Society and Culture.* Berkeley: University of California Press (1999).

28. M. Pallares-Burke (Ed.), *The New History.* Cambridge: Polity Press (2002).

29. P. Burke, *What is Cultural History?* Cambridge: Polity Press (2004).

30. E. A. Clark, *History, Theory, Text: Historians and the Linguistic Turn.* Cambridge: Harvard University Press (2004).

31. G. M. Spiegel (Ed.), *Practicing History. New Directions in Historical Writing after the Linguistic Turn.* Rewriting Histories Series, London & New York: Routledge (2005).

32. A. Green and K. Troup (Eds.), *The Houses of History: a Critical Reader in Twentieth-century History and Theory.* Manchester: Manchester University Press (1999).

33. K. Jenkins and A. Munslow (Eds.), *The Nature of History Reader.* London & New York: Routledge (2004).

34. B. Trigger, *A History of Archaeological Thought.* Cambridge: Cambridge University Press (1989).

35. I. Hodder, M. Shanks and A. Alexandrini, *Interpreting Archaeology: Finding Meaning in the Past.* London & New York: Routledge (1997).

36. G. Halsal, Archaeology and Historiography. In: M. Bentley (Ed.), *A Companion to Historiography.* London & New York: Routledge, 805–827 (2001).

37. T. Bennett, *New Keywords: A Revised Vocabulary of Culture and Society.* London: Blackwell (1999).

38. H. Ritter, *Dictionary of Concepts in History.* New York: Greenwood Press (1986).

39. K. Popper, *The Poverty of Historicism.* Boston: Beacon Press (1957).

40. K. Popper, *Conjectures and Refutations. The Growth of Scientific Knowledge.* London: Routledge & Kegan Paul (1963).

41. A. C. Danto, *Analytic Philosophy of History.* Cambridge: Cambridge University Press (1965).

42. W. Dray (Ed.), *Philosophical Analysis and History.* New York: Harper & Row (1966).

43. P. L. Gardiner (Ed.), *The Philosophy of History.* Oxford: Oxford University Press (1974).

44. F. R. Ankersmit and H. Kellner (Eds.), *A New Philosophy of History.* Chicago: University of Chicago Press (1995).

45. R. F. Berkhofer, *Beyond the Great Story: History as Text and Discourse.* Cambridge: Harvard University Press (1997).

46. W. Dray, Philosophy and Historiography. In: M. Bentley (Ed.), *A Companion to Historiography.* London & New York: Routledge, 763–782 (1998).

47. Q. Skinner (Ed.), *The Return of Grand Theory in the Human Sciences.* Oxford: Oxford University Press (1985).

48. M. Fulbrook, *Historical Theory: Ways of Imagining the Past.* London & New York: Routledge (2002).

49. M. L. Arduini, *Trattato di Metodologia della ricerca storica.* Milano: Jaca Book (1996).

50. Ch. Langlois and Ch. Seignobos, *Introduction aux études historiques.* Paris: Librairie Hachette (1898).

51. G. J. Garraghan, *A Guide to Historical Method.* Westport, CT: Greenwood Press (1973).

52. R. J. Shafer, *A Guide to Historical Method.* Wales: The Dorsey Press (1974).

53. G. R. Elton, *The Practice of History.* London: Fontana Press (1967).

54. A. Marwick, *The Nature of History.* London: Macmillan (1970).

55. G. Himmelfarb, *The New History and the Old.* Cambridge: Harvard University Press (1987).

56. G. R. Elton, *Return to Essentials. Some Reflections on the Present State of History.* Cambridge: Cambridge University Press (1991).

57. J. Appleby, L. Hunt and M. Jacob, *Telling the Thruth about History.* New York: Norton (1994).

58. R. E. Evans, *In Defence of History.* London: Granta (1997).

59. E. Fox-Genovese and E. Lasch-Quinn (Eds.), *Reconstructing history: the emergence of a new historical society.* London & New York: Routledge (1999).

60. A. Marwick, *The New Nature of History: Knowledge, Evidence, Language.* Houndmills: Palgrave (2001).

61. K. Jenkins (Ed.), *The Postmodern History Reader.* London & New York: Routledge (1997).

62. B. Fay, Ph. Pomper and R. T. Vann (Eds.), *History and Theory: Contemporary Readings.* London: Blackwell (1998).

63. W. Prevenier and M. Howell, *Reliable Sources: An Introduction to Historical Methods.* Ithaka: Cornell University Press (2001).

64. M. Stanford, *A Companion to History.* Oxford: Blackwell (1994).

65. J. Tosh, *The Pursuit of History: Aims, Methods and New Directions in the Study of Modern History.* London: Longman (1984).

66. E. Breisach, *Historiography: Ancient, Medieval, and Modern.* Chicago: Chicago University Press (1995).

67. F. R. Ankersmit, *Historical Representation.* Stanford: Stanford University Press (2001).

68. L. Jordanova, *History in Practice.* London: Arnold (2000).

69. G. G. Iggers, *Historiography in the twentieth century: from scientific objectivity to the postmodern challenge.* London: Wesleyan University Press (1997).

70. A. Munslow, *The New History.* London: Longman (2003).

71. T. Kuhn, *The Structure of Scientific Revolutions.* Cambridge: Cambridge University Press (1962).

72. J. J. Venter, Reality as History: The Historic Turn in Western Thought. Paper presented at *The Twentieth World Congress of Philosophy.* Boston, August 10–15, 1998. `http://www.bu.edu/wcp/Papers/Mode/ModeVent.htm` (1998).

73. T. J. McDonald, *The historical turn in the human sciences.* Ann Arbor: University of Michigan Press (1996).

74. P. Hamilton, *Historicism. The new critical idiom.* New York & London: Routledge (1996).

75. W. Flinders Petrie, *Methods and Aims in Archaeology.* London & New York: Macmillan (1904).

76. A. Munslow, *The Routledge Companion to Historical Studies.* London & New York: Routledge (2000).

77. M. Foucault, *Les mots et les choses. Une archéologie des sciences humaines.* Paris: Gallimard (1966).

78. M. Foucault, *L'archéologie du savoir*. Paris: Gallimard (1969).

79. R. G. Collingwood, *The Idea of History*. Oxford: Clarendon (1946).

80. A. Munslow and R. A. Rosenstone, *Experiments in Rethinking History*. London & New York (2004).

ADDRESSING THE SUSTAINABILITY CHALLENGE BEYOND THE FACT-VALUE DICHOTOMY: A CALL FOR ENGAGED KNOWLEDGE

GERT GOEMINNE

Centre Leo Apostel for Interdisciplinary Studies
Vrije Universiteit Brussel (VUB), Belgium
E-mail: gert.goeminne@vub.ac.be

FILIP KOLEN

Centre for Critical Philosophy
Ghent University (UGent), Belgium
E-mail: filip.kolen@UGent.be

ERIK PAREDIS

Centre for Sustainable Development
Ghent University (UGent), Belgium
E-mail: erik.paredis@UGent.be

In our culture, deliberation over a sustainable future takes place in a political arena with several players: politics; science, business, civil society. The debate is heavily weighted upon by the relation between science and politics: science is armed with powerful, but unconcerned facts underpinning all of the important decisions, while politics is left with nothing but values, full of concern but quite powerless. Opposed to the image of the arena, we appeal to the idea of the agora as a meeting place where experts and lay people participate in the same practice of 'engaged knowledge'. Forwarding the agora as a political space to address sustainability issues beyond the fact–value dichotomy involves three main tasks. First of all, the conceptual space needs to be opened up to host the agora. Thinking in terms of the co-constitution of subject and object and a conception of science as a necessarily engaged activity clears the way to dwell on the 'in-between' of the subject–object poles rather than always already falling back to either of both sides, be it science/policy, fact/value or knowledge/power. Secondly, we have to make clear what politics looks like in the agora. Here, participation gets a different — 'engaged' — status: engaged knowledge practitioners, lay and expert, participate in the same practice: what they do or say is accountable to a measure constituted from within the dynamics of practice, in response to what they recognize as what is at issue. Thirdly, the agora needs to be put at work. Therefore, concrete policy approaches in the broad field of sustainable socio-technical transitions are looked into as potential entry points for our agora approach of engaged knowledge.

1. Preamble

This article originates in a cross-disciplinary trialogue between the three authors: Gert Goeminne, PhD in nuclear physics and since a few years engaged in research on the science–policy interface;[1] philosopher Filip Kolen who focuses on the co-constitution of subjectivity and objectivity;[2] and sustainability scholar Erik Paredis who conducts policy-oriented research on socio-technical innovations and transitions.[3] Rather than being a classical research paper, this article aims at opening up a space for reflection and discussion on the entanglement between scientific and societal engagement. Our 'call for engaged knowledge' presented here grew out of a dual concern shared by the three authors. First of all, being scientific researchers ourselves, we are deeply concerned by an apparent lack of both intellectual space and a sense of necessity within academic research circles to deal with issues of human well-being. In the academic world, the focus is on becoming competent in a particular field of research, competency being measured by the number of publications. In this context, fundamental issues of human existence are regarded as 'soft' questions as compared to the 'hard' research questions of science. Under the ideal of 'freedom of research', the non-interestedness in social relevance is even regarded as a necessary condition for good science, claiming that the benevolent societal impact is something that will eventually happen, as a necessity in the linear chain leading from science over innovation to societal benefit, but that the researcher should not be concerned with.[a] As Langdon Winner laments in his wonderful essay *Brandy, cigars, and human values*,[4] issues of human well-being only become a theme for the scientist "as a hasty afterthought late in the day". In this paper, Winner marvellously depicts the scene of a typical late night 'reflexive panel' on an academic conference: "Out comes the brandy. Out come the cigars. And out comes the after-dinner speaker, an old trooper, usually a distinguished scientist or engineer ... to address his colleagues on 'technology and human values'." While those who engage in those discussions truly believe they are getting down to the most fundamental issues in human existence, the whole framing of such reflexive panels

---

[a] As our own crossdisciplinary experience has learned us, it is important to mention that this argument does not only hold for the natural sciences. Likewise, research in the human sciences often delivers answers to questions of which nobody seems to bother whether these are interesting, relevant or even ethical. It is also in the sense of this juxtaposition of 'research for its own sake' and 'contemplating the societal and/or ethical aspects of the outcomes of such research' that our characterization of 'hard' versus 'soft' has to be understood.

shows that these 'soft' questions are actually addressed as an add-on, as something you only bother about when the 'hard' work of science is done.

This disquieting observation is in our view mirrored by an analogue pattern of thought within policy circles concerned with intertwined issues of human development and environmental degradation. Intensive experience with policy preparatory research in such sustainability issues (see e.g. Ref. 5) has repeatedly shown us that whenever policy turns itself to science, this is almost always governed by a quest for a unequivocal framing of problems as well as for unequivocal solutions to these problems. Results of policy-oriented research should be typically 'useful' for policy, where useful has to be interpreted as 'readily implementable' and with a guarantee that 'objectives' will be achieved when 'guidelines' are followed. All this is motivated by the same one-dimensional frame of thought linking science and society already shown to be present in scientific circles and which is here interpreted as science being the only trustworthy basis for sound policy-making.

It is crucial to see that both our observations, in science as well as in policy circles, are characterized by a similar attitude of irresponsibility. The irresponsibility at the side of scientific engagement could be characterized through the following device: "Focus on producing correct answers (and publications) and do not bother about what could be important questions". The irresponsibility at the side of political engagement is rather governed by an attitude of submission to scientific certainty: "Sound policy and decision-making should solely be based on scientific evidence". As a result of this mutually reinforcing irresponsibility at both sides, a genuine, responsible discussion on what a sustainable future could be and how it could be attained is totally removed from view.

It is this fundamental concern that set the scope for the cross-disciplinary trialogue underlying the present paper. The overarching aim is to forge a way out of the baleful dichotomous framework of science and politics that has installed itself as a science-over-politics priority in Western societies. How can we get the search for a sustainable future back in view of our scientific and political endeavours without falling back on a simplistic 'progress' discourse of scientific knowledge necessarily and evidently leading to human well-being? How are we to think about science and politics, about fact and value and their interrelations in order to open up a genuine perspective on responsibility? The ambition here is to begin exploring these questions by sketching the contours of a frame of thought and action that thinks beyond simplistic demarcations between science and politics, fact and value. In a first step however, we will further elaborate on what constitutes in our view the core of the problem we want to address here.

2. The Sustainability Impasse in the Modern Constitution

We live in an age where our culture is faced with deeply intertwined global environmental problems and human development issues such as climate change and poverty. In the early nineties, initiated by the worldwide appeal of the World Commission on Environment and Development report *Our Common Future*,[6] the paradigm of sustainable development was put forward as an attempt to bridge the gap between environmental concerns about the increasingly evident ecological consequences of human activities and socio-political concerns about human development issues. While the broad goals have been widely embraced and sustainable development is now a much quoted policy objective, its implementation seems further away than ever: the cooperative global environmental governance regime envisioned at the 1992 Earth Summit in Rio is still in an institutional incubator while neo-liberal economic globalization has become fully operational with environmental degradation and global poverty still burgeoning (see e.g. Refs. 7–8).

It is our view that the crux of this impasse lies in sustainability politics finding itself stuck with its self-proclaimed ideal of rational decision-making. Environmental problems, often generated by the products of science, are indifferently framed in techno-scientific terms leading to a simultaneous scientisation and de-politicisation of environmental governance. This expert focused technological determinism, embedded in a discourse of ecological modernisation, acts to marginalise the issues of human choice involved in putting sustainability into effect and to downplay deliberation over the socio-cultural practices, behaviours and structures such choice involves.[9] As a result of this techno-scientific focus, the need for accordant social change is removed from view, which makes sustainability all the less likely to occur in practice. This is convincingly illustrated by the current impasse on climate change that has been created and maintained by making political acting subordinate to a scientific framing of what is in essence a societal problem. The narrow scientific focus on global climate change addresses itself to an undifferentiated global 'we' and relies exclusively on the authority of science to create a sense of urgency for radical change.[10] In the absence of some other basis of appeal, 'we' are likely to act as uninvolved spectators, i.e. we are likely not to act, rather than as participants in the shaping of our future, making responsible, sustainable change all the more improbable to occur.

In order to break out of this deadlock, it is first of all necessary to dig deeper into the philosophical foundations of this scientisation of sustainability politics. In our view this problematic is indeed rooted in the metaphysical and ontological options that are dominant in our Western society and that are based on a frame of thought in which subjectivity and objectivity are seen as mutually exclusive. Objectivity, on the one hand, is what science strives for; it is the realm in which truth is to be situated. Either our knowledge accurately represents reality — and it is true — or it does not, and is thus rejected. Subjectivity, on the other hand, relates to mere personal opinions (preferences, values, interests) and can never yield reliable knowledge. It is this strict ontological division between the non-human object and the human subject and its corresponding epistemology of represented facts and values (representationalism) that Latour has called the 'Modern Constitution'.[11] The political order of the Modern Constitution now takes the form of an arena with two adversaries: the first one, called science and armed with powerful, but unconcerned facts taking all of the important decisions, and the other, called politics, left with nothing but values, full of concern but quite powerless.[12] Scientific facts are the only trustworthy basis to ground political decisions leaving no room for subjective values. As already hinted at in the preamble as one of our major concerns, this ideal of rational decision-making boils down to an ideal of irresponsibility: science dismisses politics of taking decisions. In brief, this is the mainstream account of the relation between objectivity and subjectivity, which confines our scientific culture in an 'objectivism' and the well-known 'ideology of science' accompanying it.

Attempts to counter this objectivistic dogmatism in epistemological as well as in political thinking have always reasoned from within this mutually exclusive fact–value framework of the Modern Constitution. As one of the most prominent critics of this 'objectivism' in the representational epistemology of the Modern Constitution, social constructivism has concentrated on the sociological history of scientific knowledge production. Once the dissociation of subjectivity and objectivity has been (conceptually) established however, any effort to thematise the subjective embeddedness of objectivity comes too late. Any a posteriori conceptualization of subjective perspectives or social contexts inevitably leads to a threatening of the objectivity at stake, and gives rise to relativism.

A similar evolution can be observed in political thinking, especially in the context of sustainability. Starting from the idea that sustainability governance is emerging as an increasingly scientised and technocratic domain, increased citizen deliberation and participation in the scientific realm have been proposed to reverse these technocratic features. A call for increased citizen participation was also strongly voiced at the 2002 Earth Summit in Johannesburg. Recognizing uncertainty and value-ladenness as elementary characteristics of sustainability problems, new forms of developing sustainability policies through broadened public participation have been proposed. 'Post-normal science'[13] and 'sustainability science'[14] are, besides others, terms used to indicate a transition towards a new method for dealing with increased uncertainty in complex sustainability issues by 're-injecting' norms and values into scientific practice. Thinking from within the Modern Constitution however, making room for subjectivity by promoting public influence on conventional rational decision-making processes can only be thought of as a threatening of the objectivity at stake.[15] Since they still adhere to the ideal of a complete fact-based knowledge of the problem under question, these participatory approaches easily lead to a marginalisation of lay perspectives.[16] At best, this 're-injection' of norms and values into scientific practice results in a negotiated sorting out of competing interests with scientific truth as ultimate executioner where participation merely serves to secure the legitimacy and acceptability of the decisions taken.

3. Breaking the Impasse: Call for Engaged Knowledge

To sum up our account of what we are inclined to call the 'sustainability impasse': the major sustainability problems our contemporary culture is faced with ask for a profound and radical rethinking of society but our potential to do so is curtailed by the hegemony of science in our two-tiered world of fact and value. Neither fact-centred technocratic nor value-centred participatory approaches yield the space for engaged decisions and responsible action. Moreover, any approach that thinks from within this Modern Constitution is bound to end up as a symptomatic affirmation of science's hegemony since it intrinsically still adheres to the ideal of an absolute objectivity. A responsible and fruitful approach can thus only be thought of and developed beyond the fact–value dichotomy. Here, we want to think beyond this dichotomy. The overarching aim of our call for engaged knowledge is dual, namely to open up a philosophical thinking space as well as a scientific-political acting space in which sustainability challenges can be addressed in a responsible and adequate manner. We will proceed in three steps, going from a philosophical to an implementational level.[b] In a first step, we will sketch the contours of an epistemology of engaged knowledge based on an ontology in which subjectivity and objectivity do not have to appear as mutually exclusive counterparts but are seen as connected in a relation of co-constitution. Secondly, we will further explore what this co-constitution account and its conception of science as a necessarily engaged activity implies for the relation between science and politics. This will enable us to develop the conceptual instruments to open up a new perspective on the science–policy nexus in the context of sustainability. In a third and last step, we will discuss concrete policy approaches that show affinity with this new perspective and indicate how they could further benefit from an engaged knowledge approach. Here, focus will be on socio-technical sustainability transitions and the socio-cultural practices, behaviours and structures such changes involve rather than on either fact-centred technocratic or value-centred 'democratic' fixes. These three consecutive steps are further explained below.

3.1. *The Co-Constitution of Subjectivity and Objectivity: Opening Up a Conceptual Space to Think beyond the Fact–Value Dichotomy*

The Modern Constitution is distinguished by the epistemic ideal of absolute objectivity (objectivism) and a bottomless fact–value divide is its necessary condition of possibility. Objectivity in its absolute sense can only exist when it renders truth, and truth has to reflect — represent — an independent reality. Therefore, the realm of facts has to be purified from subjective relativities in order to be susceptible to truth, and an epistemic view of a strong detachment of fact and value has to be installed. In complex knowledge domains where problems are characterized by a deep intertwinement of facts and values, objectivity does not suffice to supply the answers and has to be supplemented by subjective concerns. But within this regime, every plea for more attention for subjective values turns out to be nothing

---

[b]Although we choose to develop our argument in a linear way, we want to emphasize it is characteristic of our viewpoint that the different levels are deeply intertwined and that theory (philosophy, science) has no priority over practice (policy). The intrinsic entanglement of 'hard' theoretical work and 'soft' implementional work will figure prominently throughout the further elaboration of this call for engaged knowledge, questioning once again the culturally installed distinction between 'hard' and 'soft' research questions identified in section 1.

but a further endorsement of the fact–value dichotomy and hence objectivism. Even subjectivism, notwithstanding its denial of the possibility of (absolute) objectivity and its flirting with relativism, falls back on the same scheme. Even there, the ideal of absolute objectivity, by proclaiming its impossibility, is preserved. The problem within this condition is that neither the emphasis on 'hard' facts, nor the accentuation of 'soft' values does provide an epistemic account of responsible knowledge and action.

On the one hand, science, in so far as it yields true knowledge, represents the facts disconnected from the initial problems, methods and questions through which the scientists were intentionally involved with their research objects. The normativity at work in the choices (abstraction and standardization, formal tools, etc.) the scientists make are lost in the final result. What is more, this loss of normativity is 'essential' for the possibility of objectivity — and thus 'necessary'. As such, objective knowledge is ethical neutral (or at least, that is the ideal). The validity of the answers of scientific knowledge is self-sufficient and universal, disconnected from the questions and the particular perspective of the scientists. This brings us back to the brandy and cigars of Langdon Winner (see section 1): ethical issues can be addressed as side effects of scientific knowledge (or afterwards when it is applied), but on no account these issues are able to harm its validity. As a consequence, the intrinsic engagement of the scientific enterprise lies outside the scope of Modern Constitution's basic epistemological options. The political counterpart of this epistemology, with its ideal of rational decision-making, inherits this impossibility of making the constitutive normativity thematic. Making decisions in the name of science is making decisions in no name at all.

On the other hand, any attempt at covering this deficit by supplementing scientific facts with subjective values witnesses of the same impotence. Against the background of the hegemonic universality of science, the particularity of human values cannot but appear as a matter of either arbitrary or dogmatic preference, i.e. at this side of the coin decisions are made either in the name of number and statistics or in the name of one or another God. In any case, the issue of responsible knowledge and action escapes Modern Constitution's conceptual scheme once again. Of course, the internal engagement at work in knowledge production can a posteriori be (scientifically) studied, but it is thereby externalized and objectified in terms of (social, psychological, historical) structures of power. Such studies may yield interesting information about the constitutive nature of this engagement, but at the same time they undermine the objectivity of the knowledge in question, i.e. its objective meaning is reduced to the meaning of the distribution of power relations. That is indeed the necessary outcome of the Modern Constitution: not able to address adequately the intrinsic engagement of knowledge, the only option for epistemology is to think of the decision field as an arena of struggle and of positions only concerned about their power.

All this shows the need for an epistemological account of knowledge that avoids these pitfalls. Traditionally, truth is the validity criterion of scientific knowledge. Truth is science's measure. Because the concept of truth has the connotation of disconnectedness and of representationalism (true knowledge has to be a 'mirror of nature'; see e.g. Ref. 17), we will elaborate on the less philosophical loaded notion of 'measure' as a criterion for the adequacy of scientific knowledge. We will develop this concept in such a way that it is competent to be a measure of the answers science provides (the validity of the knowledge as such) as well as of the questions that are being asked (the subjective grounds of this knowledge). Our conception of measure is embedded in a global theory of knowledge based on the concept of co-constitution. This concept is a dynamical evolution and actualization of Edmund Husserl's transcendental-phenomenological notion of constitution.[18] In short, Husserl starts from the idea that the objective world is not given as such (as it is in the representationalist account) but that it is the result of a process of transcendental constitution. Using his method of 'epoche' (suspension), Husserl discovers an absolute pole of subjectivity, the transcendental Ego, as the absolute ground of this constitution. In order to arrive at a framework that is adequate to deal with complex societal issues such as sustainability, this theory of constitution has to be radicalized. Both the objective world and the pole of subjectivity have to be 'bracketed' or 'suspended' to open up a space of thought where the mutual interrelatedness of subjectivity and objectivity appears as primal, while the relating poles are but derived. Objectivity is then no longer seen as originating from the autonomous activity of an a priori given subjectivity, but both subjectivity and objectivity are mutually co-constituted in the process that 'discriminates' them.[19] The resulting objectivity is the sedimentation of a subjective engagement, while, reversely, subjectivity only comes into being as the implicit engagement that the objective sedimentation is witnessing of. Subjectivity and objectivity are each other's necessary condition of possibility: subjectivity is the transcendental, i.e. 'juridical', condition of the objective sedimentation, while objectivity is the 'material' condition of the accompanying subjectivity.

We are convinced that a powerful and enabling epistemological account of engaged knowledge together with its intrinsic notion of measure may be built around the mathematical concept of symmetry.[20] Recent developments in (classical) philosophy of science increasingly pay attention to the notion of symmetry as a valuable instrument in their field.[21] Most (if not all) of them interpret symmetry as a merely heuristic tool practicable at the surface of a framework that is basically that of the Modern Constitution. However, the symmetry idea is much more powerful when it is seen as an implementation of the co-constitution view on objectivity. The symmetry-view on co-constitution is inspired by the success of the use of mathematical symmetries in modern physics. The elegance of symmetry principles is usually seen as an indication for the ultimate objectivity of the theories built around them. In our interpretation, however, the use of symmetries not so much illustrates the objectivity of these theories, as that they show what it is to be objective. Beyond being objective, they illustrate the constitution of the theories' objectivity. The basic idea behind it is very simple, but the extraordinary epistemological consequences when it is implemented as a theory of co-constitution, still have to be explored in full detail. Briefly, the idea is the following. Symmetries are defined as invariants under transformations of coordinates. The objective poles of sedimentation can be identified with the invariants that are involved in symmetries. This means that something can be objective only relative to a set of operations, to a set of transformations of coordinates. These coordinates can be connected with a certain perspective, and the transformations with the notion of subjective engagement. While in traditional accounts of objectivity 'relativity', 'perspective' and 'grounding in subjectivity' are threatening the objectivity at stake, in the symmetry view they are, on the contrary, constitutive for it.[20]

The symmetry view on objectivity is now intrinsically tied up with an ethics of science that is proper to the essence of scientific practice, and that neither just deals with deontological rules nor with the a posteriori application of scientific results. While this intrinsic normativity fell out of scope of Modern Constitution's epistemology, it is situated in the heart of the co-constitution view, because of the latter's focus on the necessary interrelatedness of subjectivity and objectivity. Subjectivity is conceptualized as the internal necessity of the external constituted objectivity. As a consequence, scientific knowledge is essentially engaged knowledge. In light of this normativity and engagement, the question of measure can now be adequately addressed. Our philosophical framework of engaged knowl-

210

edge is competent to avoid the 'irresponsible' measure of either (too) global objectivity or (too) local subjectivity. On the one hand, the measure of engaged knowledge is not just (objective) knowledge. It is not grounded in a disconnected, self-sufficient foundation. The conceptual framework of co-constitution guarantees the impossibility of an absolute foundation, and the internal necessity of a constitutive engagement. The subject has to carry the measure and is therefore responsible for it. On the other hand, the measure of engaged knowledge is not just (subjective) engagement. The measure is not reduced to the local preferences and perspectival choices of subjectivity. According to the symmetry view, the engagement consists in the commitment of the subject to install a certain indifference with respect to its actual perspective (without transcending it in 'a God's eye view'). The necessary relativity of the own perspective and the openness to others is a condition for the possibility of making responsible decisions, i.e. of decisions that are not just arbitrary choices. The measure of co-constitution is essentially a measure of responsibility.

3.2. *From Arena to Agora: Developing the Conceptual and Institutional Contours to Rethink the Science–Policy Nexus beyond the Modern Constitution*[c]

By forwarding the idea of co-constitution and the related conception of science as an essentially engaged activity, we are now enabled to open up the conceptual space where simplistic demarcations between science and policy, facts and values, knowledge and power can be critically assessed and challenged. By focusing on the entanglement and the way the sedimentations of co-constitution crystallize out, we aim to lighten up the 'in-between' of the subject–object poles rather than always already falling back to either of both sides. As the world and its problems do not present themselves in scientific terms, an important starting point is to see how matters of concern are — through science — always already framed into matters of fact.[5] According to our co-constitution view, scientific matters of fact such as CO_2-concentrations or viruses are the objective sedimentations of a subjective (i.e. scientific) engagement, while subjectivity (e.g. the scientific norms and preferences) comes into being only as the engagement of that sedimentation. As outlined above, this co-constitution process involves, of necessity, implicit matters of judgment, priority, choice and interpretation

---

[c]This paragraph is partially based on Ref. 1.

(in symmetry jargon: installing a certain indifference with respect to its actual perspective) as the transcendental conditions of the objective sedimentation, the latter often embodied in explicit paradigmatic scientific procedures of abstraction and standardization (See e.g. Refs. 1 and 19). This scientific framing results in a model of the system of interest couched in terms of 'matters of fact'. In the Modern Constitution, this model is disconnected from its genesis and its validity gets a universal status: it is regarded as a true representation of the problem at stake and serves as the only valid basis possible for policy-making. Considered in the context of its construction, however, a model aims to fulfil a certain function, and the choice of function depends on what kind of knowledge is aimed at and what the model is supposed to account for and to take into account: a problem is not chosen or given, it is formulated, framed, given shape.[22] Latour uses the notion of 'circulating reference' to emphasize the aspects of reduction and amplification involved in the 'in-between': the representational view of the world conveyed by such models renounces many of the key attributes of the original context, such as materiality, particularity and locality (reduction), while at the same time amplifying key attributes of matters of fact such as compatibility, standardization and text.[23] This analysis suggests that if we wish to liberate ourselves from the Modern Constitution and therefore be in a better position to facilitate sustainability, we require a non-representational epistemology illuminating these reduction and amplification aspects of the 'in-between' and an appreciation of how knowledge does not so much reflect a pre-given state of the world but acts to shape it in ways that both facilitate and constrain action. As outlined in section 3.1, the symmetry-view on co-constitution serves as a powerful starting-point in elaborating such an epistemology.

Rather than conceiving of knowledge in terms of representations of the (social or material) world the co-constitution perspective and its concept of engaged knowledge emphasizes the socio-material practices from and within which these representations arise. Such a view then also changes the way we think about ourselves and our place in the world in fundamental ways: the world now becomes something that we are embedded in and part of rather than being detached from and merely observers of, as representationalism suggests. Whereas the latter tends to paint the 'natural world' as an enduring material backdrop to and constraint on our activities, the concept of engaged knowledge, recognizing the complex ways in which our practices shape the world, reveals the many potential configurations it may take, dependent on our choices and the practices to which these give rise. In this respect, Actor-Network-Theory developed by Latour, Callon and Law could be employed to illuminate the way knowledge producing and deploying practices configure and reconfigure networks of relations in ways that both enable and constrain human and non-human actions and interactions (For a critical 'reflective' reflection on Actor-Network-Theory see Ref. 24). This brought Latour to the conclusion that science is indeed politics pursued by other means.[11] Indeed, an epistemology of engaged knowledge, developed along the lines sketched above, can make convincingly clear how the practice of producing and applying knowledge is inherently linked to political thinking and acting or how — following Jasanoff — natural and social order are co-produced.[25] It allows for instance to look for new concentrations of power in places that are hardly ever considered in current social theory, such as the causally constructed climate models which are heavily relied upon in global climate policy.[10]

A crucial step, which we can only begin to explore here, now consists in employing the insights derived from our non-representational account of engaged knowledge in the elaboration of the institutional contours to facilitate social change for sustainability. A full appreciation of how knowledge acts to shape the world in ways that both enable and constrain choice over the futures available gives us two inherently related starting points: one related to the issue of participation, the other related to the concept of measure with regard to the questions of science introduced above. Firstly, it has already been argued how the Modern Constitution acts to obscure how collaborations between scientists and non-scientists, by categorising the latter's practical and behavioural insights as mere preferences or values, might put lay insights at work in adequately addressing sustainability issues. Therefore, if sustainability is to be about human choice, and thus about the different practices and behaviours such choices embody rather than about specific technological options and trajectories, then engaging lay people in these terms in addressing the problem at stake is an imperative.[9] The second starting point builds on the insight that the major concern of representationalism and its — vain — attempts to increase citizen participation in the scientific realm, is focused on the justification of the answers science provides; that is on the efficacy of whichever material or social factors are regarded as legitimating the representation of concern. Of primary concern to the practice-focused account of engaged knowledge provided here however, is rather an enduring concern with the questions science poses and with matters of adequacy and relevance.[d] Inspired by the ideas on the impossibility of an absolute foundation of the measure of engaged knowledge introduced above, true political questions now take centre stage: What is at issue?; To whom and to what does it matter? and How can it appropriately and adequately be described?[26]

Building on these two starting points, we suggest the term 'engaged participation' to denote this form of participation that on the one hand facilitates collective understandings, embodying a diversity of insights, and on the other hand, is reflexively occupied with the question of measure. In our account of objectivity, lay knowledge is genuinely integrated, not by categorizing it as mere values or preferences and in this way paralyzing them, but by focusing on their role in the 'in-between' of the constitution of objectivity. Analogously, expert knowledge is no longer paralyzed as matters of fact by granting them an absolute status. Both expert and lay knowledge are called upon to participate in addressing relevant issues and what connects them is an enduring concern with the question of measure. Here, participation gets a different — engaged — status. Engaged knowledge practitioners, lay and expert, participate in the same practice in that what they do or say, is accountable to a measure constituted from within, in the dynamics of practice, in response to what they recognize as what is at issue in the practice. As opposed to the image of the arena, as the political order of the Modern Constitution where paralyzed facts fight and win over paralyzed values due to the cultural superiority of the former in our scientific culture, we propose the image of the agora as the political order of the co-constitution ontology. In its broadest sense, the agora, referring to the central place in ancient Greek cities where genuine discussion took place, denotes a 'working space' where engaged participation is made possible and engaged knowledge is at work. In a more particular sense, we want to explore how recent approaches in sustainability policy and science could benefit from our conceptual agora framework.

---

[d]With respect to our practice-focused account of knowledge, Joseph Rouse's work on practice theory constitutes an important reference in further elaborating on our concept of measure (see e.g. Ref. 26). In developing a 'normative' conception of (scientific) practices as constituted by the mutual accountability of their performances, Rouse understands normativity in terms of accountability to what is at issue and at stake in a practice. Although we touch upon this connection with Rouse here, one of the authors elaborates in more detail in Ref. 27 on the parallels that can be drawn between the concept of measure and Rouse's conception of normativity.

3.3. *Putting Engaged Knowledge at Work in Science–Policy Agorae for Sustainability Transitions*

Political science often makes a distinction between three policy paradigms: a top-down steering model with a central role for government and planning instruments, a bottom-up market model with a large degree of autonomy for societal actors and use of market instruments, and a governance model with networks of actors searching for and deliberating over shared rules and agreements through learning processes. Although a lot of studies on sustainability politics plead for a shift to a governance model, actually existing sustainability policies are usually a combination of planning (e.g. national plans for sustainable development, regulation) and market approaches (e.g. taxes, subsidies, tradable permits). Even though we are by now more than 20 years after the Brundtland report *Our Common Future*[6] and more than 15 years after the UNCED conference in Rio, these approaches show hardly any signs of being able to initiate a profound restructuring of developed countries' economies towards more equity and sustainability. A lot of countries have some form of sustainable development plan, but these cannot make the difference; neither seem market instruments able to fundamentally reorient production structures and consumption patterns. In an attempt at scientifically objectifying the debate and in that way putting sustainability policies higher on the agenda, from the mid nineties onwards the sustainability community has put a lot of effort in the development of new measuring instruments, indicators and indices. The indicator culture that has grown over the years predominantly fits in a rational-instrumental approach to decision-making: a well-chosen and scientifically grounded indicator informs the policy-maker, who can then take the correct decisions and guide policies towards more sustainability. As has been argued above, this kind of approach to science and policy traps sustainability policies in a fact–value dichotomy and marginalizes deliberation over a fundamental restructuring of society.

However, over the last decade, a new generation of sustainability policy and science seems to be growing that may offer a hybrid framework at the science–policy interface to address radical change through a more reflexive and action-oriented approach. We refer in particular to the literature and practices around socio-technical system innovations and transitions (See e.g. Refs. 28–33). In the transition perspective, current industrialised societies are analysed as societies wrestling with complex, persistent problems in the systems that are central for their welfare creation, such as the food system, the energy system, the mobility system and the health system. Radical innovations at the level of these systems are deemed a prerequisite for changes towards more sustainable development in industrialised countries. A transition then is a transformation process in which existing structures, institutions, culture and practices are broken down and new ones are established. From studying historic transitions — such as those from horse-drawn carriages to the automobile, or from sailing ships to steamships[31] — it has become clear that transitions usually take a long time to develop (25 to 50 years) and go through different stages, that they take place through interactions between different levels of a system (niche, regime, landscape), that they are not caused by a single factor but by an interplay of many factors that influence each other (technological, economical, ecological, socio-cultural), and that they are the result of interaction between networks of actors. In other words, they are multi-actor, multi-factor, multi-level and stretching over a long period of time.[30]

Transition researchers explicitly state that radically reorienting societal systems towards sustainable development cannot be done with traditional policy and science recipes. The first reason is that the problems we are struggling with are deeply rooted in our institutions, structures, practices, patterns of thinking and value systems. Traditional policy and science approaches merely perpetuate these structures and modes of thinking. Secondly, the problems are highly complex and perceived differently by different actors, so that problem definitions as well as favoured solutions differ. This undermines the idea that one rational scientific explanation can be found that can unambiguously advise policy. Third, the future development of problems is highly uncertain, also opposing the idea of non-contradictory policy guidelines based on sound science. The consequence is that, starting from the complexity, interdependency and uncertainty that are characteristic of our society, new policy- or governance-approaches need to be developed which take into account the inherent conflict of interest, opinion and value in sustainability politics.[33] Since sustainability cannot be unequivocally defined, the transitions approach advises not only to focus on the direction development should take, but also to pay more attention to the processes leading into that direction. Grin, referring to Charles Lindblom, pleads for a mode of knowledge generation called 'inquiry' or 'probing'.[34] This is a process in which experts and other actors, based on different kinds of knowledge (scientific, experiential, local, tacit) try cooperatively to define a problem and find solutions, in the process also discovering values, preferences, power issues that are relevant for the problem at hand. The agora concept, introduced above as a working space where engaged participation is made possible and engaged knowledge is at work, now provides the contours within which such an 'inquiry' and 'probing' may take place. Indeed, the agora is constituted as the place where lay and expert knowledge are on equal footing through their common accountability to the question of measure in response to what they recognize as what is at issue.

Not unsurprisingly, the transition perspective struggles with translating theoretical insights into practical guidelines for policy development. In what is currently the mainstream version of the governance approach that is developed, called 'transition management', innovators with various backgrounds, perspectives and ambitions are brought together to develop a common problem perception of the targeted social system (e.g the energy system), develop a long-term sustainability vision for that system, pathways towards the vision and short-term experiments that are meant to test the pathways in practice.[e] The hope is that through social learning in this network of innovators and through the development of shared long-term visions and agendas, it is possible to influence the expectations and actions of societal actors, and to increase influence on regular short-term policies. However, the practical application of transition management in Dutch policy has shown that it runs the risk of obscuring its own politics, "smoothing over conflict and inequality; working with tacit assumptions of consensus and expecting far more than participatory processes can ever deliver."[35] In other words, although transition researchers are aware of the flaws of the Modern Constitution, representationalism can still slip in through the backdoor. The vision formulated above on the science–policy nexus and built around the idea of the agora as the place where engaged knowledge is at work, opens up several research tracks that may be able to strengthen the system innovation and transition approach to governance for sustainable development.

First, the role of science and the scientist in processes of socio-technical system innovation and transition management has to be made a theme of research. While this is, perhaps surprisingly, hardly done at the moment, reflecting on the relation between science, policy and social change through transition processes will be a necessity to overcome the trap of the Modern Constitution. Following Grin,[34] we can state that if transition and

---

[e]In the light of the argument developed in this article, it is ironic that the term used to describe the place where these actors meet, is 'transition arena'. Perhaps a 'transition agora' would have been more suited.

system innovation researchers are to contribute to developing governance approaches that do not suffer from the flaws of the classical-modernist idea, it is crucial that they themselves avoid these traps. This demands a thorough reflection in these processes on the role of science and the scientist, the way these interact with other actors and other forms of knowledge, the implications for the status of the knowledge generated, its potential impact for social change *etcetera*. In other words, the way in which engaged knowledge can be conceptualized in transition studies is an important future research goal.

Second, even when these topics can be clarified, this does not automatically mean that the insights can directly be translated into guidance of how to practically position science and the scientist in these processes. This is important, though, because if it is not possible to formulate an actual practice of engaged knowledge and participation, the yielded insights remain impotent vis-a-vis real-life problems of sustainable development and systems in transition. More in particular, a series of questions pose themselves in relation to the implications of an agora where engaged knowledge and participation for transitions is practiced. Under which conditions does engaged knowledge and participation become possible in a transition process? How can such processes be set up? How can the relevant actors, their positions and knowledge be taken into account? What are the roles of the different actors in these processes? Which forms will (social) learning take? Is it possible to pronounce upon the quality of process and results? Can criteria be defined for repeating a comparable process in a different context? Reflecting on these type of questions and making progress in answering them constitutes another crucial goal in the study of science–policy agorae for transitions in sustainable development.

4. Conclusion

This call for engaged knowledge is first and foremost a call for responsibility in shaping our society. We have indeed argued that it is timely and necessary to move beyond a mutually reinforcing pattern of irresponsibilty that is present between scientific and political spheres. In elaborating the concept of engaged knowledge, we have shown that this irresponsibility is the very result of a fundamental characteristic of our Modern Constitution, namely that of systematically separating scientific facts and technological instruments from values, choices and responsibilities. Arguing that natural and social order are rather co-produced as a result of the inherent entan-

218

glement between scientific and societal engagement, our call touches upon three interconnected levels. On a conceptual level, we focus on the entanglement between object and subject, knowing and acting and nature and society. On a socio-political level, the entanglements between knowledge and power and science and politics take centre stage. And finally, on an implementational level, focus is on the entanglements between social and technological change for sustainability. Throughout our analysis we have hinted at several areas for further research and action and it is our hope that these may attract 'engaged' knowledge practitioners around this call. Opening up this paper for discussion constitutes a first step in this direction. In this respect and in congruence with the central argument of this paper, we willingly acknowledge the political character of this paper.

Acknowledgments

Gert Goeminne's research underlying this paper has been made possible through a Postdoctoral Fellowship of the Research Foundation – Flanders. Erik Paredis acknowledges support of the Policy Research Centre for Sustainable Development of the Flemish Government. The authors thank Michel Lavrauw for his generous encouragement and helpful comments on an earlier version of this paper.

References

1. G. Goeminne, Once upon a time I was a nuclear physicist. What the politics of sustainability can learn from the nuclear laboratory. *Perspectives on Science* (In print).
2. F. Kolen and G. Van de Vijver, De terugkeer van het verdrongene? Een analyse van de subjectieve gronden van objectieve kennis naar aanleiding van van Fraassens The Empirical Stance. *Tijdschrift voor Wijsbegeerte* **70** (2), 317–338 (2008).
3. E. Paredis, W. De Jonge, E. Vander Putten, F. Maes, J. Lavrysen and S. Van Passel, Vlaanderen in transitie? In: M. Van Steertegem (Ed.), *Toekomstverkenning MIRA-S 2009*. Mechelen: Vlaamse Milieu Maatschappij (In print).
4. L. Winner, Brandy, cigars, and human values. In: L. Winner, *The Whale and the Reactor*. Chicago: The University of Chicago Press, 155–163 (1986).
5. G. Goeminne and E. Paredis, The concept of ecological debt: Challenging established science-policy frameworks in the transition to sustainable development. In: E. Techera (Ed.), *Frontiers of Environment and Citizenship*. Oxford: Inter-Disciplinary Press (In print).
6. G. Brundtland (Ed.), *Our Common Future: The World Commission on Environment and Development*. Oxford: Oxford University Press (1987).

7. C. Sneddon, R. Howarth and R. Norgaard, Sustainable development in a post-Brundtland world. *Ecological Economics* **57**, 253–268 (2006).

8. W. Lafferty and J. Meadowcroft, *Implementing sustainable development: strategies and intitiatives in high consumptive societies.* New York: Oxford University Press (2000).

9. S. Healy, Epistemological Pluralism and the 'Politics of Choice'. *Futures* **35**, 689–701 (2003).

10. D. Demeritt, The Construction of Global Warming and the Politics of Science. *Annals of the Association of American Geographers* **91**, 307–337 (2001).

11. B. Latour, *We Have Never Been Modern.* Cambridge, MA: Harvard University Press (1993).

12. B. Latour, Ein Ding ist ein Thing. *Concepts and Transformations* **1** (2), 97–111 (1998).

13. S. Funtowicz and J. Ravetz, Science for the Post-Normal Age. *Futures* **25**, 735–755 (1993).

14. R. W. Kates et al., Sustainability Science. *Science* **27**, 641–642 (2001).

15. G. Goeminne, On the need and (im-)possibility of a sustainability science. *Proceedings of the XXII World Congress of Philosophy, 2008, Seoul, Korea* (In print).

16. B. Wynne, May the sheep safely graze? A reflexive view of the expert-lay knowledge divide. In: S. Lash, B. Szerszynski and B. Wynne (Eds.), *Risk, Environment and Modernity: Towards A New Ecology.* London: Sage, 44–83 (1996).

17. R. Rorty, *Philosophy and the mirror of nature.* Princeton, NJ: Princeton University Press (1980).

18. E. Husserl, *Ideas pertaining to a pure phenomenology and to a phenomenological philosophy. 2: Studies in the phenomenology of constitution.* Translated by R. Rojcewicz and A. Schuwer. Dordrecht: Kluwer (1989).

19. F. Kolen, The processes of objectification in science, art and everyday life as inherently engaged activities in a social world. In G. Lasker, H. Schinzel and K. Boullart (Eds.), *Art and Science.* Windsor Ontario, Canada: IIAS, **2**, 49–53 (2005).

20. F. Kolen, An interpretation of Husserl's concept of constitution in terms of symmetry. *Analecta Husserliana. Volume 88. Logos of Phenomenology and Phenomenology of the Logos.* Dordrecht: Kluwer, 307–317 (2005).

21. T. Debs and M. Redhead, *Objectivity, invariance and convention: Symmetry in physical science.* Cambridge, MA: Harvard University Press (2007).

22. I. Peschard, Participation of the Public in Science: Towards a New Kind of Scientific Practice. *Human Affairs* **17**, 138–153 (2007).

23. B. Latour, *Pandora's Hope: Essays on the Reality of Science Studies.* Cambridge, MA: Harvard University Press (1999).

24. B. Latour, *Reassambling the Social. An Introduction to Actor-Network-Theory.* Oxford: Oxford University Press (2005).

25. S. Jasanoff (Ed.), *States of Knowledge: The co-production of science and social order.* London: Routledge (2004).

26. J. Rouse, *Engaging Science: How to Understand Its Practices Philosophically*. New York: Cornell University Press (1996).
27. G. Goeminne, Postphenomenology and the Normativity of the In-between: The Politics of Sustainable Technology. *Foundations of Science* (accepted for publication).
28. A. Rip and R. Kemp, Technological Change. In: S. Rayner and E. Malone (Eds.), *Human Choice and Climate Change*. Columbus, Ohio: Batelle Press, 327–399 (1998).
29. J. Rotmans, R. Kemp and M. van Asselt, More evolution than revolution: transition management in public policy. *Foresight* **3** (1), 15–31 (2001).
30. B. Elzen, F. Geels and K. Green (Eds.), *System Innovation and the Transition to Sustainability. Theory, Evidence and Policy*. Cheltenham, UK and Northampton, MA: Edward Elgar (2004).
31. F. Geels, *Technological Transition and System Innovations. A Co-Evolutionary and Socio-Technical Analysis*. Cheltenham, UK and Northampton, MA: Edward Elgar (2005).
32. A. Smith, A. Stirling and F. Berkhout, The governance of sustainable socio-technical transition. *Research Policy* **34**, 1491–1510 (2005).
33. D. Loorbach, *Transition Management, new mode of governance for sustainable development*. Utrecht: International Books (2007).
34. J. Grin, The multilevel perspective and design of system innovations. In: J. van den Bergh and F. Bruinsma (Eds.), *Managing the Transition to Renewable Energy. Theory and Practice from Local, Regional and Macro Perspectives*. Cheltenham, UK and Northampton, MA: Edward Elgar, 47–79 (2008).
35. E. Shove and G. Walker, Caution! transitions ahead: politics, practice, and sustainable transition management. *Environment and Planning A* **39**, 763–770 (2007).